M I A s

A Reference Handbook

Other Titles in ABC-CLIO's
CONTEMPORARY
WORLD ISSUES
Series

Abortion, Second Edition	Marie Costa
Biodiversity	Anne Becher
Campaign and Election Reform	Glenn Utter and Ruth Ann Strickland
Children's Rights	Beverly C. Edmonds and William R. Fernekes
Crime in America	Jennifer L. Durham
Domestic Violence	Margi Laird McCue
Drug Abuse in Society	Geraldine Woods
Endangered Species	Clifford J. Sherry
Environmental Justice	David E. Newton
Euthanasia	Martha Gorman and Carolyn Roberts
Feminism	Judith Harlan
Gangs	Karen L. Kinnear
Legalized Gambling, Second Edition	William N. Thompson
Militias in America	Neil A. Hamilton
Native American Issues	William N. Thompson
The New Information Revolution	Martin K. Gay
Recycling in America, Second Edition	Debra L. Strong
Religion in the Schools	James John Jurinski
School Violence	Deborah L. Kopka
Sexual Harassment, Second Edition	Lynne Eisaguirre
United States Immigration	E. Willard Miller and Ruby M. Miller
Victims' Rights	Leigh Glenn
Violence and the Media	David E. Newton
Violent Children	Karen L. Kinnear
Welfare Reform	Mary Ellen Hombs
Wilderness Preservation	Kenneth A. Rosenberg
Women in the Third World	Karen L. Kinnear

Books in the Contemporary World Issues series address vital issues in today's society such as terrorism, sexual harassment, homelessness, AIDS, gambling, animal rights, and air pollution. Written by professional writers, scholars, and nonacademic experts, these books are authoritative, clearly written, up-to-date, and objective. They provide a good starting point for research by high school and college students, scholars, and general readers, as well as by legislators, businesspeople, activists, and others.

Each book, carefully organized and easy to use, contains an overview of the subject; a detailed chronology; biographical sketches; facts and data and/or documents and other primary-source material; a directory of organizations and agencies; annotated lists of print and nonprint resources; a glossary; and an index.

Readers of books in the Contemporary World Issues series will find the information they need in order to better understand the social, political, environmental, and economic issues facing the world today.

M I A s

A Reference Handbook

Jeanne M. Lesinski

CONTEMPORARY WORLD ISSUES

ABC-CLIO

Santa Barbara, California
Denver, Colorado
Oxford, England

Library of Congress Cataloging-in-Publication Data

Lesinski, Jeanne M.
    MIAs : a reference handbook / Jeanne M. Lesinski.
        p.   cm. — (Contemporary world issues)
    Includes bibliographical references and index.
    ISBN 0-87436-954-1 (alk. paper)
    1. Vietnamese Conflict, 1961–1975—Missing in action—United
States—Handbooks, manuals, etc. I. Title. II. Series.
UB803.L47   1998
959.704'38—dc21   98-42275
                                                                          CIP

03 02 01 00 99 98   10 9 8 7 6 5 4 3 2 1

ABC-CLIO, Inc.
130 Cremona Drive, P.O. Box 1911
Santa Barbara, California 93116-1911

This book is printed on acid-free paper ∞.
Manufactured in the United States of America

In memory of my father

Contents

List of Tables, xiii
Preface, xv

1 Introduction, 1
    The Issue, 1
    Definitions, 2
    The Lists, 2
    Historical Overview, 4
    The Vietnam Conflict, 6
    Casualties and Treatment of
       POWs, 11
    The MIA Issue after Vietnam, 15
        Ending the War, 15
        The Defense Intelligence Agency,
           17
        Families and Activists Demand
           Accounting, 20
           Effects on Families, 20
           Private Efforts, 21
           Hoaxes, 21
        Congressional Activity, 24
        Presidential Perspectives, 26
    Today's Debate on Vietnam
       MIAs, 30
    Political Aftermath, 35

Looking Ahead, 39
References, 40

2 Chronology, 45

3 Biographical Sketches, 59

4 Statistics and Documents, 69
Statistics, 70
Documents, 74
  Code of Conduct for Members of the Armed Forces of
    the United States, 75
  Tap Code, 76
  Personal Accounts of POWs and Their Families, 76
    "*Escape and My Darkest Hour,* " 76
    Excerpts from the Testimony of Carol Hrdlicka, 91
    Excerpts from the Statement of Judy Coady Rainey, 96
    Excerpts from the Testimony of Patrick J. Cressmen,
      100
    Excerpts from the Testimony of Colonel Theodore
      Guy, 106
Speeches and Quotes, 113
  Position on United States Relations with Vietnam, 26
    January 1994, 113
  Winston Lord's Speech on Lifting the Trade
    Embargo, 9 February 1994, 117
  Bob Dole's Speech against Normalization,
    10 July 1995, 130
  President Clinton's Speech on Normalizing
    Relations, 11 July 1995, 133
  Proclamation 6818: National POW/MIA
    Recognition Day, 1995, 135
Treaties and Legislation, 137
  The Geneva Convention for the Protection of War
    Victims: Armed Forces in the Field (1949), 137
  Paris Peace Accords (1973), 147
  Complete Text of the Letter from President
    Richard Nixon to Prime Minister Pham Van Dong,
    1 February 1973, 152
  National Defense Authorization Act, Section 569
    on Missing Persons, 154

The POW/MIA Flag, 157
References, 159

5 Directory of Organizations, 161

6 Selected Print Resources, 193
General Reference Works, 193
Books, 196
Government Reports, 205
Periodical, 208

7 Selected Nonprint Resources, 209
Audio Book, 209
Computer Programs, 209
Internet Sites, 211
Radio Talk Show, 217
Videotapes, 218

Glossary, 225

Index, 231

List of Tables

Table 4.1. American Casualties and Prisoners, 70

Table 4.2. Americans Unaccounted for in Southeast Asia, 70

Table 4.3. Listing by State of Unaccounted-for Servicemen, 71

Table 4.4. Americans Accounted for since 1974, 73

Preface

The MIA—missing in action—issue is a complex one, particularly because international politics and classified information make inquiries and understanding difficult. Moreover, all wars ruin lives and shatter families, so an emotional component is added to the issue. The Vietnam conflict is no exception. For Americans the Vietnam conflict of the 1960s and early 1970s turned into a no-win situation, a negotiated peace in which the United States as a nation lost lives and prestige. Although many American prisoners of war were returned following the peace accords, questions have long remained as to whether the Communist victors kept prisoners and the remains of deceased military personnel to use as bargaining chips in their efforts to extract reparations from the U.S. government. Other questions remain concerning the resolve of the U.S. government to fully account for the missing. In the wake of the 1974 Watergate scandal and President Richard Nixon's subsequent resignation, many citizens no longer believed that they could trust their public officials.

The families of those missing in action desire to know the fate of their loved ones and have in some cases spent years and substantial sums of money in their efforts to learn the truth. They and their supporters have brought to the issue the determination and zeal of religious crusaders. Over the decades, the federal government of the United States, in conjunction with Vietnamese, Cambodian, and Laotian officials, has attempted with varying degrees of effort and success to resolve the MIA issue.

This book provides a balanced survey of the resources available and a guide to further research on the topic of prisoners of war (POWs) and soldiers missing in action during the Vietnam conflict. Chapter 2 is a chronology of events pertaining to the conflicts in Southeast Asia, the release of prisoners of war, and the activities of governmental bodies, public and private organizations, and individuals trying to resolve the issue. Chapter 3 provides biographical sketches of some figures who have been involved in the issue in significant ways. Chapter 4 includes statistics on casualties and prisoners, documents related to the treatment of prisoners, congressional testimony, and a map of POW camps in Vietnam. Chapter 5 is a listing of private organizations involved in the issue. Chapters 6 and 7 provide bibliographies of print, nonprint, and electronic sources of further information.

Introduction 1

The Issue

The Vietnam conflict was the first war after which the U.S. government was compelled to account for all of its military personnel. Although some individuals believe this accounting is a moral imperative, others maintain that it is an overly expensive and unnecessary proposition that has prevented the development of productive relations with Vietnam and the resolution of problems plaguing Vietnam veterans.

Among those active in the MIA issue are the families of prisoners of war, congressional representatives, interested citizens, and the employees of the relevant government agencies. Each party has its own agenda, and these goals may conflict, generating further controversy about the issue.

Two opposing camps exist: those who believe that live POWs continue to be held captive in Southeast Asia, and others who maintain that no prisoner could have survived captivity to this day. Among the former camp, some believe that the Vietnamese also continue to warehouse the remains of deceased soldiers and are holding back critical intelligence records that could resolve many of the MIA cases.

Many family members of MIAs believe

they have been misled and mistreated by the U.S. government and that the relevant government agencies have been ineffective in addressing their concerns, including live-sighting reports. Yet the U.S. government has continually stressed its commitment to as full an accounting as possible. On the basis of what he considered to be significant progress on resolving the MIA issue, President Bill Clinton normalized relations with Vietnam in 1995 despite the strenuous objections of families of MIAs and others.

Definitions

As in many specialized fields, jargon is commonly used in the military. Therefore, it is necessary to define the terms used to discuss casualties. During the Vietnam conflict (note that it was never declared a "war" by Congress and so technically remains a "conflict") casualties were defined as "any person who is lost to the organization by reasons of having been declared dead, missing, captured, wounded, injured, or seriously ill" (U.S. Department of Defense, *U.S. Casualties*, 1985). A person could be listed as killed in action (KIA), killed in action/body not recovered (KIA/BNR), prisoner of war (POW), or missing in action (MIA). Of all these categories, missing in action is the most nebulous, for the fate of the person could not be readily determined. Listed as MIA are "active duty military personnel who are not present at their duty station due to apparent involuntary reasons and whose location is not known" (U.S. Department of Defense, *U.S. Casualties*, 1985).

The Lists

In the case of POWs and MIAs from the Vietnam conflict, different numbers have been generated at different times (see Chapter 4). One reason for the fluctuating numbers is that different government agencies used varying methods to compile their lists (Keating 1994, 21). From 1963 to 1973, the Pentagon listed POWs and MIAs separately. They were only combined into an MIA category after the Paris Peace Accords had been signed in January 1973. Of the approximately 2,000 servicemen unaccounted for at war's end, half had been originally listed as KIA/BNR; these servicemen were known to be dead, but their bodies could not be recovered for whatever reason (for example, they were incinerated during the in-air explosion of an aircraft) (Franklin 1992, 96–98;

McConnell 1995, 30–31). A squadron commander in Vietnam related how the process of determining casualties favors declaring a person missing rather than dead because the person whose duty it is to make that determination is usually a friend of the serviceman in question, and after a serviceman is declared dead his spouse no longer receives his salary and benefits, just a lump-sum death benefit (Clarke 1979, 37–44; Keating 1994, 22–23). Although the U.S. Air Force denied that a policy of deliberate deception on status existed, a retired Air Force lieutenant colonel recounted that the unofficial policy in his search-and-rescue squadron was for the air crewmen to say if they would rather be listed MIA or KIA/BNR if they could not be rescued after being shot down (McConnell 1995, 70–71).

Additionally, after Operation Homecoming—the name given to the negotiated release of the 653 American prisoners of war who returned to the United States en masse during the period from 27 January to 4 April 1973—deserters from the theater of operations were added to the MIA category, though the official definition of missing excludes those "who are in an absent-without-leave or deserter status or those who have been dropped from the rolls of military service" (U.S. Department of Defense, *U.S. Casualties*, 1985). It had always been known by the military that at the Vietnam conflict's end there were "scores of AWOLs and deserters, including escaped stockade and brig prisoners, loose in the criminal world 'twilight zones' of South Vietnam's cities" (McConnell 1995, 140). Robert Garwood is the most famous of the deserters from the war, although after his return to the United States in 1979 desertion charges were disproved. Susan Katz Keating (1994, 26–27) cited four other cases of desertion that came to light in the late 1980s and early 1990s. Moreover, the Department of Defense denied the existence of defectors, those who aided the enemy, during the war because it did not want the Vietcong to profit from them for propaganda purposes; yet as many as 40 MIAs may actually be defectors (Keating 1994, 27).

Prisoners who were repatriated at various times prior to the 1973 Operation Homecoming served as human memory banks. They memorized prisoners' names and data about them and provided that information to the U.S. government. One such prisoner was Doug Hegdhal, whose phenomenal memory supplied the government with important information about 260 prisoners (Norman 1990). Other repatriated prisoners may not have been as accurate as Hegdhal, who knew of prisoners from firsthand experience; they sometimes remembered names of men rumored to be prisoners but who were already dead. Furthermore, the operatives of the Central Intelligence Agency (CIA) who were involved in covert operations in Cambodia and Laos and were killed were often purposely listed as missing from Vietnam (Keating 1994, 21–22). Malcolm McConnell (1995, 33) pointed out that POW statistics have been adversely affected by "painfully inadequate intelligence on the number and location of U.S. POWs held throughout Indochina," despite enemy prisoner and defector debriefings and other intelligence efforts.

Over the years, the Department of Defense (DoD) has made public various listings of personnel declared POW or MIA. In the 1990s, it went on the Internet with a monthly update of this list at its Defense Prisoner of War/Missing in Action Office Web site (see Chapter 7). Though believed to be deceased, Charles Shelton was for many years symbolically listed as the only remaining U.S. POW of the Vietnam conflict (U.S. Department of Defense, *Factbook,* 1990, 9), and in 1993 his children asked the U.S. government to change that designation to KIA (Keating 1994, 245). All others who were at one time thought to be POWs have been declared KIA, KIA/BNR, or MIA. In 1977, the Carter administration ordered that scores of men be reclassified from MIA to KIA/BNR, so that by 1978 there were only 224 official MIAs. This action caused an uproar among some of the families of those servicemen on the list. Through Congress and the courts, the families forced the reinstatement of those names and the addition of some 1,200 known to be dead to the MIA list in 1980 for further investigation (Keating 1994, 27–28). Since the signing of the Paris Peace Accords in 1973, remains have been repatriated and identified at varying rates (see Chapter 4). As of 1 April 1998, 2,093 American servicemen remained on the list of those unaccounted for. However, it is important to understand that this number really reflects the number of servicemen for whom no remains have been recovered; although a photograph or other evidence may indisputably prove the death of an individual, under congressional mandate a name cannot be removed from the list unless a live person or identified remains are recovered.

Historical Overview

In ancient times, vanquishers killed their enemies outright rather than take them prisoner, in large part because they could barely maintain themselves above a subsistence level, let alone feed and shelter their captives. Though ancient Greeks believed themselves to be imbued with human dignity, they killed or sold into slavery their prisoners. The Romans tortured captives, used them for public combat against gladiators, or enslaved them (Barker 1974). During the European Middle Ages, when a code of chivalry governed warfare, knights would spare the lives of other knights, instead holding them hostage and demanding a ransom. Foot soldiers, however, fared less well—they faced extermination by their captors (Mowery, Hutchings, and Rowland 1968).

Humanism arose during the seventeenth and eighteenth centuries; this prompted greater efforts to control warfare. The seventeenth-century Dutch jurist Hugo Grotius proposed that wars should only be fought for "just" causes. Although the rules he wrote to change the activities of belligerents were not adopted, they influenced later thinkers. During the 1700s, such French philosophers as Charles de Montesquieu and Jean-Jacques Rousseau and the Dutch jurist Emmerick de Vattel proposed that captors should treat prisoners humanely, safeguarding them, because the conflict was one between nations, not individuals. During the American Revolution, the British considered the colonial freedom fighters to be criminals, not prisoners of war, and executed them or held them in such unhealthy sanitary conditions that they died of disease or starvation (Barker 1974).

Nearly a century later, little had changed in the treatment of prisoners. During the American Civil War, the North and South held prisoners in such deplorable conditions that the public called for reform. At prison camps, particularly Florence and Andersonville in the South and Johnson's Island and Point Lookout in the North, prisoners lacked medical care, lived in overcrowded unsanitary conditions, and suffered from diseases and malnutrition (U.S. Department of Defense, *POW*, 1955). In response to public outcry, President Abraham Lincoln issued U.S. War Department General Order 100, "Instruction for the Government of Armies of the United States in the Field," otherwise known as the Lieber Code for its author, Columbia College professor Francis Lieber. Among its provisions, the code stipulated that prisoners should not be declared criminals, tortured, starved, or killed. Lieber maintained that they are prisoners of the government, not their captors, and must be held in safety until the end of hostilities.

An international code for the treatment of prisoners of war was developed during the late nineteenth century and early twentieth century. It began with the Geneva (Switzerland) Convention of 1864, which marked the founding of the International Red

Cross organization and set down the "Convention for Amelioration of the Wounded in Time of War." This document stipulated
that medical personnel and facilities should remain unscathed
during wars and that combatants and civilians should be treated
alike despite their political positions. A decade later the Russian
government convened a conference in Brussels, Belgium, during
which participants discussed a code similar to that of Lieber. Although the representatives of the assembled nations did not ratify
a code, the precedent had been set for similar conferences.

In 1899, 1907, and 1914, national leaders met in The Hague,
Netherlands, for peace conferences, and in 1929 a convention
took place in Geneva. The resultant "Convention Relating to the
Treatment of Prisoners of War" mandated that captors treat prisoners humanely, supply them with information, and allow visits
by neutrals, such as Red Cross workers. Of the 46 nations represented at the conference, 33 agreed to the convention, but abuses
of the convention by Germany during World War II necessitated
another Geneva Convention in 1949 to renew a commitment to
the humane treatment of prisoners. (See Chapter 4 for applicable
portions of the conventions.)

The Vietnam Conflict

The territory currently recognized as Vietnam has been the object
of conquest by various peoples for centuries. For a thousand
years it was ruled by China. In the mid–eighteenth century, Indochina (which included what became Vietnam, Laos, and Cambodia) was captured by the French and became a French colony.
During World War II, the Japanese controlled Vietnam. When
Japan surrendered to Allied forces in August 1945, the Communist Viet Minh, led by Ho Chi Minh, seized Hanoi, the capital city
of Vietnam, and forced Emperor Bao Dai to abdicate. Ho Chi
Minh became president of the new state, the Republic of Vietnam, commonly known as North Vietnam. Although the French
colonial government recognized the new republic, the French
had reclaimed its colonies in the South, and the two governments
could not agree on economic and political matters. War erupted
in December 1946. On 1 July 1949, Emperor Bao Dai established
Saigon (now Ho Chi Minh City) as capital of South Vietnam.

The roots of the Vietnam conflict as Americans know it began
in 1954 at the Geneva Conference, which ended the Indochina
War (1946–1954) between Vietnamese nationalist and Communist

forces and the French colonial government under Bao Dai and its supporters. Representatives from France, Great Britain, the United States, China, Cambodia, Laos, the Vietnamese Bao Dai government (the French colonial government), and the Viet Minh government (Vietnamese Independence League) under Ho Chi Minh participated in this conference. An armistice between fighting French forces and the Viet Minh was reached. The Bao Dai government, which was supported by the U.S. government, did not sign the agreement, though it tacitly agreed to it. After the French troops withdrew from Vietnam, which had been divided along the seventeenth parallel of latitude into two zones separated by a demilitarized zone, the Ho Chi Minh government consolidated its power over the North. Many North Vietnamese Catholics were anti-Communists and fled to the South; at the same time, thousands of southerners went north because they believed in Ho Chi Minh. They traveled along what became known as the Ho Chi Minh Trail through Laos.

In 1950, U.S. President Harry S Truman formally recognized South Vietnam. Truman and later Presidents Dwight D. Eisenhower and John F. Kennedy considered a democratic Vietnam to be critical in preventing a takeover by the Communists of other Southeast Asian countries. This presumed cascading of regimes later became known as the "domino theory." In October 1954, President Eisenhower offered South Vietnam direct economic assistance and sent U.S. military advisors to train the South Vietnamese in the use of American weapons. The U.S. government supported the non-Communist ruler of South Vietnam whether it was the Emperor Bao Dai or Ngo Dinh Diem, a popular nationalist leader who became premier of South Vietnam and in 1955 deposed Bao Dai in a referendum. Diem proclaimed South Vietnam a republic with himself as president. He also refused to hold the reunification elections that had been mandated at the Geneva Conference because he, with the support of the United States, maintained that fair, free elections could not take place due to widespread fraud.

Fighting began in January 1957 when the Viet Minh launched guerrilla attacks on the South. The insurgents tried to get Vietnamese peasants to help them through emotional appeals and by promising them land. When the Diem government charged in 1960 that the Viet Minh were part of the North's attempts to conquer the South, the Hanoi government maintained that the guerrilla movement was independent. The Viet Minh then created its own political coalition of religious, nationalist, and social groups.

This National Liberation Front (NLF), which was made up of local militia, provincial units, and well-trained battalions, was also known as the Vietcong. The NLF proposed to replace the Diem regime with a democratic government, reform land ownership, maintain a neutral foreign policy in South Vietnam, and unify the country through negotiations.

The NLF used guerrilla tactics such as kidnappings, bombings, and murdering of peasant leaders and their families to create chaos in South Vietnam, thus undermining the Diem government and taking control of large areas of territory, village by village. Anticipating an attack by the entire North Vietnamese army, the Diem government created a conventional army of 100,000 men called the Army of the Republic of Vietnam (ARVN). Their armaments and tactics of conventional warfare were ineffective against the North's guerrilla tactics. After reiterating its support for the Diem government, the U.S. government under President John F. Kennedy in April 1961 signed a treaty of amity and economic relations.

In December 1961, asserting that the United States would help South Vietnam remain independent, President Kennedy ordered 400 uniformed Army personnel to Saigon to operate military helicopter companies that served as armored personnel carriers for South Vietnamese soldiers to pursue Vietcong guerrillas. Within a year, 12,000 U.S. civilian and military personnel had been sent to South Vietnam. Some of these personnel were in the Special Forces, known as the Green Berets. Because U.S. troops were supposed to train—not fight—in Vietnam, the Green Berets trained the ARVN and the mountain villagers in Laos to defend themselves. In the process, the Green Berets fended off attacks against themselves and the South Vietnamese. They also brought the villagers medical supplies and weapons.

In November 1963, the Diem government came to an end when Diem and his brother (who was also his political advisor), Ngo Dinh Nhu, were murdered. A revolutionary government under Brigadier General Duong Van Minh took control, but not for long. In the next year and a half, more coups d'état took place: The control of South Vietnam changed ten times. In 1966, Generals Nguyen Van Thieu and Nguyen Cao Ky headed a military council that restored order. Finally in 1967, General Nguyen Van Thieu was elected president of South Vietnam.

During this period of great instability, U.S. involvement in the war increased. In 1964, a North Vietnamese patrol boat allegedly attacked the U.S. Navy destroyer *Maddox* near Haiphong, a major

North Vietnamese port in the Gulf of Tonkin. As a result, the U.S. Congress gave President Lyndon B. Johnson special power to act in Southeast Asia through the so-called Tonkin Gulf Resolution. No formal declaration of war against North Vietnam was ever made by the U.S. government, a fact that the North Vietnamese used to justify their claim that American POWs were really war criminals and did not have to be treated humanely. In retaliation for the attack on the *Maddox,* President Johnson ordered 64 navy jets to bomb North Vietnamese targets such as oil-storage tanks and patrol boats. During this maneuver, the first U.S. citizen was taken prisoner, Navy Lieutenant Everett Alvarez, a pilot.

During the early years of the conflict, U.S. soldiers took defensive positions: They protected other soldiers, aircraft, and fuel depots in what is known as the "fortress strategy" while ARVN soldiers took offensive action. After the Vietcong began to attack U.S. military bases in South Vietnam in 1965, President Johnson, at the urging of General William Westmoreland, commander of U.S. forces in Vietnam, ordered American warplanes to bomb targets in North Vietnam in "Operation Rolling Thunder." This operation lasted eight years and was meant to break the will of the North Vietnamese. The United States also sent American combat soldiers to assist the ARVN officers, who were considered to be incompetent and ineffective.

Although U.S. infantry soldiers were better armed with rifles and mortars than their Vietcong enemies, they were also burdened with heavy supply packs. In contrast, the Vietcong carried few supplies and were able to move stealthily and quickly through the jungle, where they hid supplies in caches to be retrieved as needed. Aircraft of all kinds played a large role in U.S. military operations. Fighters, bombers, and reconnaissance and transport planes cluttered the skies. Helicopters served as troop carriers, air hospitals, and battlefield supply carriers.

Despite its successes in battle, the U.S.–South Vietnamese combination was rife with problems. U.S. and ARVN forces alienated the peasants when they carelessly defoliated food crops with Agent Orange, relocated entire villages and bulldozed areas looking for underground Vietcong hideouts, bombed large areas of land, treated peasants harshly while searching for Vietcong, and seemed to support the urbanites who had luxuries while the peasants had nearly nothing. Also in favor of the eventual success of the Vietcong was the seemingly infinite size of the fighting force, which replaced its killed soldiers with new ones almost immediately from a large population.

By 1967, some U.S. military advisors believed that the war in Vietnam might last much longer than the leaders in Washington had anticipated. Antiwar protest, which had begun on college campuses in 1965, continued to gain momentum and was now voiced by a wide range of Americans. Televised reports, though censored, brought the war into millions of homes, giving citizens pause to consider, and question, the decisions of their government.

Indecisive military strategy may have lost the war. When the Vietcong acted, the United States reacted—instead of taking the initiative in military actions. Many intensive battles were fought along the demilitarized zone (DMZ), which many U.S. leaders perceived of as a floodgate holding back a tide of Communists. On 31 January 1968, during a weeklong national new year's celebration called Tet, the Vietcong and North Vietnamese army launched massive, coordinated attacks on military centers in Saigon, Hue, Da Nang, and other important cities. Although U.S. and ARVN forces eventually repulsed the North's forces from the urban areas, large rural areas remained under Vietcong domination. The offensive psychologically harmed American morale both in Vietnam and in the United States.

In Vietnam, General Creighton Abrams replaced General Westmoreland as commander of U.S. forces, and shake-ups among U.S. officials took place. President Johnson declined to run for reelection and stopped the bombing of North Vietnam in an effort to bring about peace talks. The talks went nowhere. When Richard Nixon was elected president of the United States in 1968, he did so partly on the basis of his hinting that he had a plan to end the war. In January 1969, President Nixon instituted what became known as Vietnamization, that is, a gradual withdrawal of American ground troops so that the ARVN would take over most operations by 1972. U.S. military personnel continued to hold advisory and air support posts, and bombing of North Vietnam and its borders increased. Many Americans listed as missing in action were pilots shot down over Cambodia and Laos.

Illegally, President Nixon ordered U.S. troops to invade and bomb Cambodia, a supposedly neutral country that was allowing Vietcong and North Vietnamese troops to rest and unload supplies in its territory in order to invade South Vietnam. This action was terminated when the U.S. Congress voted to cut off funds for the operation. Although U.S. operatives had previously conducted covert operations in Laos using CIA operatives, in 1971 Nixon ordered U.S. troops to openly invade Laos as well, to disrupt North Vietnamese troop movements along the Ho Chi Minh Trail.

After taking heavy loses during the Tet Offensive, the Vietcong and North Vietnamese army mainly employed guerrilla tactics until April 1972, when they attacked and captured several major South Vietnamese cities, including Quang Tri. As a result, President Nixon ordered full-scale bombing and mining of Vietnamese ports and waterways. The ARVN was able to recover some of the lost territory by the early summer, and in October 1972 U.S. National Security Advisor Henry Kissinger was secretly negotiating with Le Duc Tho of North Vietnam. After much disagreement, representatives of the North and South signed a peace agreement in Paris on 23 January 1973.

Casualties and Treatment of POWs

The exact number of casualties for all parties during the Vietnam conflict is unknown. The Vietnamese militaries, North and South, did not keep accurate records of their military forces. The U.S. military, however, kept detailed personnel records, including records of those killed. As of 1985, the total deaths of U.S. military personnel numbered 58,022 (see Chapter 4). Among prisoners of war held in the Hanoi camps, 5 percent died while in captivity, compared with a 25 percent death rate for those held by the Vietcong in jungle camps in the South and in Laos (Veterans Administration 1980, 40). The disparity in death rates reflects the vast difference in conditions of urban and rural camps.

During the Vietnam conflict, two distinct categories of prisoner of war emerged: those captured in the North, who were held in Hanoi area prisons, and those captured in South Vietnam or Laos, who were held in jungle camps. To varying degrees, all prisoners of the Vietcong suffered from a lack of medical treatment, from malnutrition, and from torture. Treatment of prisoners was particularly harsh prior to 1969, when the Paris peace talks began, yet conditions improved only marginally as the talks progressed.

Northern prison camps were situated in and around Hanoi. In northwestern Hanoi the prisons included Son Tay (called Camp Hope by POWs), Briar Patch, Camp Faith, and Dan Hoi. To the southwest of Hanoi was Skid Row, and in the south was Rockpile. Hanoi proper contained an old French prison named Hao Lo (called Hanoi Hilton by POWs), Plantation Gardens, Alcatraz, the Zoo and Zoo Annex, and Dirty Bird. Of the camps, the Hilton and Plantation Gardens were the largest (Rowan 1973, 22–23). These ancient buildings lacked plumbing and were poorly ventilated.

Prisoners were often confined in large numbers in close quarters, or alternatively in solitary confinement to prevent communication between high-ranking officers and lower-ranking personnel.

Dozens of survivors later described their captivity in memoirs (Alvarez and Pitch 1989, Anton 1997, Blakey 1978, Chesley 1973, Johnson and Winebrenner 1992, McGrath 1975, Mulligan 1981, Risner 1973, Rowan 1973, Stockdale 1984, Zalin 1975). Treatment seemed to follow no prescribed formula and varied widely between prisons and among prisoners. Ted Guy, held in Plantation Gardens, described being kept in a 10-by-12-foot concrete-floored room, bare except for the wooden planks that served as a bed. Shutters covered the windows, and a dim, unshaded light-bulb in the ceiling burned constantly. Guy was provided two pairs of black pajamas, two pairs of shorts, a pair of sandals, two blankets, a mosquito net, one bar of lye soap every 45 days, a toothbrush, and a small container of toothpaste every two months. A defecation bucket was made available and emptied daily. At this location, prisoners were allowed to use the "shower room" twice each week, which meant soaping up and rinsing with a bucket of water (Zalin 1975, 252–253). Dick Stratton, in contrast, recalled being given one cup of water to wash with each week and of having to go 200 days without a shave or haircut. Under these conditions the stench of prisoners, who were often ill with dysentery, was indescribable (Rowan 1973, 25).

Prior to 1969, prisoners were fed two meals daily. Each consisted of a half-loaf of bread or a small portion of white rice and a vegetable or soup made of pumpkin, kohlrabi, cabbage, or greens. The quart of water given each prisoner daily proved to be adequate during the winter months but inadequate during the high heat and humidity of the summer months. Occasionally prisoners received bamboo shoots or peanuts, and four times during the year they received "holiday meals," which included poultry—a small chicken or turkey divided among 35 men—and a small bottle of beer or a 1-ounce cordial. Whenever it appeared that peace talks might succeed—as after 1969 with the Paris peace talks—the prisoners received three meals daily. They then were given more vegetables, condensed milk, and maybe an orange, banana, or pineapple (Rowan 1973, 19–20). Prisoners also received a daily ration of cigarettes, which nonsmokers traded for food.

Paramedics gave rudimentary medical treatment to prisoners—basic first aid for wounds, immunizations, vitamin shots, and aspirin. Most prisoners' medical needs were ignored until they became seriously ill. Prisoners who were taken to a North

Vietnamese hospital for treatment were usually returned to the camp within a day or two (Veterans Administration 1980, 40).

Allowance for exercise varied widely. Some prisoners were chained to their beds for months (Veterans Administration 1980), whereas others were permitted brief exercise periods outdoors (Zalin 1975, 264). Despite the starvation diet, many prisoners practiced regular exercise regimes to maintain muscle strength and stamina. For example, Ted Guy did push-ups and walked circles equaling 15 miles per week in his tiny cell (Zalin 1975, 256).

Indoctrination efforts went on ceaselessly by means of a camp radio. A loudspeaker in each room poured forth a constant barrage of broadcasts meant to convince prisoners of their misinformed role in the conflict. In addition to programs from the Voice of Vietnam, Hanoi's official radio station, camp-recorded tapes were aired. Lessons in Vietnamese history and the tenets of communism intermingled with prisoners' "confessions" and censored news about the course of the war. Prisoners were pressured continually to make antiwar propaganda tapes to influence both camp and world opinion about the war (Zalin 1975, 256). Often this pressure took the form of torture.

Torture was commonplace before 1969. Many prisoners were tortured immediately following their capture to extract military information or simply to prove that they could be "broken" (McConnell 1995, 48–53; Rowan 1973, 26–27). Thereafter the frequency and severity of such treatment varied with the camp director and the individuals involved. Numerous survivors reported having to endure torture (Alvarez and Pitch 1989, Blakey 1978, Chesley 1973, Johnson and Winebrenner 1992, McGrath 1975, Mulligan 1981, Risner 1973, Rowan 1973, Stockdale 1984, Zalin 1975). This treatment took the form of sleep deprivation—being made to sit for days on a wooden stool—being hung upside down, beatings using ropes to contort the body into unnatural positions, or a stint in the "hot box"—a metal shed with no ventilation (Rowan 1973, 19). Solitary confinement, often lasting months, was itself a form of torture. Dick Stratton recounted how his captors went a step further by keeping him in a sensory deprivation cell for 25 days (Rowan 1973, 66). Although U.S. military personnel were well versed in the Code of Conduct (see Chapter 4), they had not been taught resistance techniques, and the human body and spirit can only withstand so much pain. Senior-ranking officers in the various prisons ordered captives through secret communications networks (see Tap Code, Chapter 4) to resist to the best of their ability. Yet they ordered prisoners to use a fallback position, that is, to lie or reveal unimportant information, to save their lives (Rowan 1973, 42). As former POW Frank Anton stated, "The vast majority of POWs were guilty of violating the Code of Conduct. The ones who refused to give the North Vietnamese anything but name, rank, and serial number didn't come home" (Zalin 1975, 341).

In the jungle camps, conditions were so harsh that torture was unnecessary. Jungle camps were located in parts of South Vietnam under the control of the North and in Laos. They were mobile and very primitive. Vietcong forces frequently moved prisoners, who were largely foot soldiers, through the mountainous jungle terrain to avoid U.S. air attacks (Veterans Administration 1980). Prisoners sometimes were held in a thatched hut, called a hootch. In one case, the kitchen was a dirt hole with stones around it, and the latrine was a hole in the ground at one end of the compound. Eighteen men slept on one bamboo platform, using old burlap bags sewn together as blankets (Anton 1997, 35; Zalin 1975, 86–87).

Jungle prisoners lacked the necessities of medical care and food to an even greater extent than did their urban counterparts. The prisoners might have received care from a medic before arriving at the camp, but the field medics were largely incompetent. Prisoners had to cook their own food, which consisted of white rice, manioc, and greens such as banana leaves, which they had to harvest themselves with the permission of the Montagnards, tribal people living in the mountainous region of central Vietnam. Rarely did they receive any source of protein, so most suffered from dysentery and then beriberi, a life-threatening deficiency of thiamine (Zalin 1975, 120–121). After the prisoners started dying in great numbers, the Vietcong gave them larger food rations, including canned goods, fish, and pork. They also received lye soap, toothbrushes and toothpaste, mosquito nets, sleeping mats, and sandals made from old car tires (Anton 1997, 85; Zalin 1975, 177).

Prisoners in jungle camps usually escaped torture except perhaps upon their capture. Instead, many were given a two-week indoctrination course by a visiting Communist cadre. He asked the prisoners to write letters of apology to the Vietnamese people and threatened their lives if they did not do so (Zalin 1975, 130). Several prisoners from one camp were, after their return to the United States, charged with collaborating with the enemy, but the charges were later dropped. As former prisoner Ike McMillan explained, "Davis [Ike's friend] and I talked about the Code of Conduct all the time, what would happen if we did this, if we did that. I could see nothing we did to violate the code. But the Code of Conduct—it's really hard to abide by once you're under pressure. What's more important? Your life or the possibility of a court-martial when you get back? I never had the feeling I would rather die than disclose information" (Zalin 1975, 107).

One civilian pilot captured in Laos, Ernest Brace, was kept for three years in a bamboo cage so small that he could not stand. Only during his twice daily trips to the latrine could he stretch his legs. He survived the frequent visits of rats, snakes, scorpions, and biting insects. Brace was tortured after his repeated attempts to escape; he was sent to Hanoi near the end of the war (Brace 1988).

Eventually most, but not all, prisoners who survived the jungle camps were consolidated into prisons in Hanoi prior to being repatriated during Operation Homecoming. (Veterans Administration 1980, 67). Former prisoners were treated at Veteran's Administration hospitals, debriefed, and returned to their families. A long-term study monitoring the health of former U.S. Navy prisoners revealed that they suffered few permanent psychological effects. However, permanent physical problems included a high rate of shoulder injuries and degenerative diseases of the joints (Kern 1989).

The MIA Issue after Vietnam

Ending the War

Several questions have frequently arisen about the ending of U.S. involvement in the Vietnam conflict. Did President Nixon knowingly abandon prisoners of war in his haste to end hostilities, and did the Communist North Vietnamese hold back prisoners to use to extract reparations from the United States? In a speech in 1973 during which President Nixon announced that "all of our American POWs are on the way home," he indicated that the accounting for MIAs in Southeast Asia was inadequate (Senate Select Committee 1993, 8). During testimony before the Senate Select Committee in 1992, former secretaries of defense Melvin Laird, Elliot Richardson, and James R. Schlesinger testified that "there was a possibility that some Americans were left behind" (Senate Select Committee 1993, 123) and referred to 20 or 30 servicemen who may have been involved in the secret war in Laos. Henry

Kissinger objected strenuously to the suggestion that he and President Nixon had not done their best to recover all American prisoners, calling such a contention "unforgivable libel" and stating, "It is the ultimate irony that our Herculean efforts to get an accounting in 1973 should be twisted twenty years later into 'evidence' that we knew POWs had been left behind" (Lippman 1992, A6–A7). The Senate Select Committee concluded:

> Given the Committee's findings, the question arises as to whether it is fair to say that American POWs were knowingly abandoned in Southeast Asia after the war. The answer to that question is clearly no. American officials did not have certain knowledge that any specific prisoner or prisoners were being left behind. But there remains the troubling question of whether the Americans who were expected to return but did not were, as a group, shunted aside and discounted by government and population alike. The answer to that question is essentially yes. (Senate Select Committee, 1993, 7)

In a prepared statement for the Committee on National Security in June 1995, information researcher Roger Hall contended that declassified CIA reports from 1967 to 1972 "show that there were up to 60 or more U.S. POWs held in Laos during the Vietnam war that were never released" and that neither President Nixon nor the Congress were properly informed of this situation (*Accounting* 1996a, 78–84). Malcolm McConnell (1995, 18–72) pointed out that even in classified military circles, opinions on the conclusions to be drawn from the available intelligence data varied widely. Whereas one analyst believed POWs were living in Laos after 1973, another judged the intelligence agents' reports to be badly flawed and inconclusive.

Henry Kissinger described in *Years of Upheaval* (1982) his negotiations with Le Duc Tho to end the Vietnam conflict. The terms involved a reconstruction program to be funded by the United States. As early as 1969, President Nixon had offered economic aid to North Vietnam and the rest of Southeast Asia. According to Kissinger, these offers were made as a "voluntary act, not an 'obligation' to indemnify Hanoi" (39) as Le Duc Tho later maintained, and they would rely on congressional approval and the living up to of the Paris Peace Accords by the North Vietnamese. Several days after the signing of the Paris Peace Accords, the United States received a list of POWs supposedly held in Laos; and a letter from President Nixon, drafted by Kissinger and his staff, was sent to Prime Minister Pham Van Dong (see Chapter 4). This secret letter, the existence of which was denied by the U.S. government until it was declassified in 1977, "suggested the procedures for discussing economic aid" through a Joint Economic Commission in order to work out "a precise aid program" (39) involving a proposed $3.25 billion in aid. Suspecting a trick, Prime Minister Van Dong repeatedly stated that he believed that the congressional obstacles were merely a pretext for denying aid. Pretext or not, when the North Vietnamese continued hostilities after the withdrawal of U.S. forces, they breached the Paris Accords by taking over South Vietnam and invading Cambodia, thus legally forfeiting any aid that a reluctant Congress might have approved. One scholar notes:

> It is obvious that the United States did attach many political qualifications to any aid, just as it is even more obvious that the North Vietnamese violated the Paris Agreement from the start. It would appear that the Nixon letter, the peculiar codicil that Kissinger brought to Hanoi immediately after the letter was sent, and the negotiations by the American delegation to the Joint Economic Commission talks were largely window dressing—that the administration fully realized that there was virtually no possibility of any significant amount of aid to Vietnam passing the Congress—but that Nixon and Kissinger were willing to dangle the possibility of such aid in front of the Vietnamese in an attempt to obtain concessions from them on other matters. (Clarke 1979, 66)

These two matters have often been interpreted by MIA activists as indicating duplicity on both sides, with the United States reneging on offers of aid and the North Vietnamese withholding POWs in case of such an eventuality.

The Defense Intelligence Agency

The Office for Prisoners of War and Missing in Action (DIA POW/MIA Office), prior to its reorganization into the Department of Defense Prisoner of War/Missing in Action Office (DPMO) in 1993, was one of many agencies whose personnel were involved in the POW/MIA issue, though from the public's perspective it was the only organization responsible for resolving the issue. In fact, over the years the issue has been treated by a bewildering array of organizations. At war's end the Four Party Joint Military Team (FPJMT), the Joint Casualty Resolution Center (JCRC), the U.S. Army Central Identification Laboratory (CILHI), and various military groups and intelligence agencies were involved. The FPJMT and JCRC were established specifically to implement Article 8 of the Paris Peace Accords, which provided for repatriating all POWs and remains and exchanging information about MIAs.

Since 1980, efforts to resolve the POW/MIA issue have been coordinated through the POW/MIA Interagency Group (IAG), which includes representatives from the Defense Department, the White House National Security Council (NSC), the State Department, the Joint Chiefs of Staff (JCS), and the Defense Intelligence Agency (DIA). For many years the National League of Families of American Prisoners and Missing in Southeast Asia also played a role in this group, but it no longer does so (U.S. Department of Defense, *Factbook*, 1990, 13). The IAG's scope includes a wide range of POW/MIA related matters, such as intelligence activities, communications with families, diplomatic efforts, public awareness efforts, and policy toward Vietnam (Senate Select Committee 1993, 31). In 1991, the Joint Task Force–Full Accounting (JTF-FA) replaced the JCRC (Stern 1995, 73).

In 1993, the Department of Defense created the Department of Defense Prisoner of War/Missing in Action Office (DPMO) to join four disparate Department of Defense offices that had been working on the POW/MIA issue: Deputy Assistant Secretary of Defense (POW/MIA), the Defense Intelligence Agency Special Office for POW/MIAs, the Central Documentation Office, and Task Force Russia (U.S.House of Representatives 1997, 103–104). Of all the agencies involved in resolving the POW/MIA issue, the DIA POW/MIA Office has had the central role, including investigating and analyzing reports of live sightings or other evidence that prisoners are possibly being held in Southeast Asia.

The performance of the DIA POW/MIA Office during the 1970s and 1980s was much criticized by families of POWs and MIAs, representatives of Congress, and even some of its own employees. The organization underwent several internal reviews in the mid-1980s and was the subject of a congressional investigation in 1991. Families of POWs and MIAs often complained that they were not given accurate information, that they were not kept up-to-date, and that the DIA did not respond to inquiries under the Freedom of Information Act. Furthermore, the perception that some DIA analysts operated from a mind-set to debunk became prevalent, and disputes about the analysis of certain intelligence information led some critics to believe that some analysts were incompetent (Cawthorne 1991, Jensen-Stevenson and Stevenson 1990, Sauter and Sanders 1993, and family testimony, Chapter 4). Activists in the POW/ MIA issue also proposed conspiracies and cover-ups to explain the DIA's lack of success in resolving cases (Cawthorne, 1991, Defense Intelligence Agency 1991, O'Daniel 1979, Sauter and Sanders 1993). However, the following year a task force headed by Lt. General Eugene F. Tighe Jr. found "no evidence of a cover-up by DIA" (Senate Select Committee 1993, 55) but did recommend a complete overhaul of the DIA (McConnell 1995, 138).

In what became a much publicized event, Colonel Millard A. Peck resigned as chief of the DIA POW/MIA Office after serving in that position for less than a year. He complained of a mind-set to debunk among DIA operatives, of inordinate emphasis being placed on the return of remains over a search for live prisoners, the fallacy that the matter is of the highest national priority, and the misuse of DIA analysts' time and expertise for what amounted to "busywork" (*DIA* 1991, 9–23). The following year a task force headed by DIA Director Kimball Gaines concluded that it had "no confidence that the current analytical process has adequately addressed all relevant factors and has drawn totally reliable conclusions" (Senate Select Committee, 1993, 55).

Onetime chief of the DIA POW/MIA Office Joseph A. Schlatter Jr. (1987–1990) pointed out that the scope of efforts at the DIA is determined by the emphasis put on it by U.S. policy and by the resources allocated by the administration. In 1979, fewer than ten employees staffed the DIA POW/MIA Office. According to Schlatter, in 1987 the office was still severely hampered by insufficient resources—both human and physical—and was too impacted by outside influences, such as requests for information from congressmen, that prevented analysts from focusing on collection and analyzing information relevant to the cases under investigation. Yet he maintained, "The facts do not support the criticism of a mindset to debunk" (*Oversight Hearings* 1994, 121–123). Likewise, the Senate Select Committee determined that the DIA POW/MIA Office was "plagued by a lack of resources; guilty of over-classification; defensive toward criticism, handicapped by poor coordination with other elements of the intelligence community; slow to follow up on live-sighting and other reports; and frequently distracted from its basic mission by the need to respond to outside pressures and requests." Several committee members asserted that "on occasion, individuals within DIA have been evasive, unresponsive and disturbingly incorrect and cavalier," but other members noted that some ˙individuals within the DIA POW/MIA Office have "performed their work with great professionalism and under extraordinarily difficult circumstances both at home and abroad" (Senate Select Committee, 1993, 20).

In response to these internal and external investigations, the Department of Defense upgraded the DIA POW/MIA's physical facilities, hired more personnel, and established closer working relationships with other agencies in the intelligence community. After reorganization of the office into the DPMO in 1993, General James W. Wold, its director, made special efforts to update POW/MIA family members through regularly scheduled briefings and newsletters (*Accounting* 1996a, 240–241; see Chapter 7).

Families and Activists Demand Accounting

Effects on Families

The negative effects due to the uncertain loss of a husband, father, or son who is listed as MIA are complex and vary from family to family. In the early 1970s, Hamilton McCubbin and colleagues studied the various emotional, social, and legal problems encountered by families of POWs who were later repatriated and those who continued to be listed as MIA. Living in a constant state of uncertainty—limbo—takes a toll on emotional and physical health. Children were forced to grow up in the absence of their fathers, and like wives and parents, they were often unable to work through their grief. "In one way or another, we families have become emotional cripples," stated the wife of an MIA, who had been active in the National League of Families of American Prisoners and Missing in Southeast Asia since its creation in 1970. "I feel that some families are so distraught and frustrated and angry that they no longer look for an accounting, but are waiting for a resurrection" (*Missing Persons* 1976b, 20).

Rod Colvin (1987) provided a forum for a dozen families of MIAs to describe their suffering, and Nigel Cawthorne (1991) recounted the stories of some of those who still wait for a missing man to return. Congressional hearings have also given the families of MIAs a forum (*Americans Missing* 1976b, *Missing Persons* 1976, *Oversight Hearings* 1994a). In 1995, Colleen Shine, whose father is listed as MIA, told Congress that she has been involved in the issue since she was eight years old (*Accounting* 1996a, 99). Although many families have grieved the missing as dead and gone on to lead their lives, others have not been able to find closure. One extreme example is Marian Shelton, whose husband was lost in action in 1965. After years of trying to determine her husband's fate, Shelton committed suicide in 1990 (Keating 1994, xx).

Private Efforts

Families of MIAs, hired case researchers, activists, and members of the media have investigated individual MIA cases by appealing to members of government agencies and to their congressional representatives, using the Freedom of Information Act to gain access to documents, making trips to Southeast Asia to search for clues, and testifying before congressional panels. In many cases, the families have spent years and significant sums of money in their efforts (Colvin 1987, Jensen-Stevenson and Stevenson 1990, Sauter and Sanders 1993, *Accounting* 1996a).

Some efforts have involved operations aimed at physically rescuing or recovering information concerning possible American POWs. These include Operation Skyhook II in the early 1980s to find prisoners in Laos, an effort in 1988 to do the same, the Team Falcon operation in 1991–1992, and the efforts of retired Army Lieutenant Colonel Bo Gritz. None of the operations has been successful in turning up a live POW (Senate Select Committee 1993, 34), and some may have been attempts to use the issue for private gain (McConnell 1995, 174, 177–178). An intelligence operation that did make a significant breakthrough was initiated privately by former United Nations worker and university librarian Ted Schweitzer, who was allowed access to the hidden archives at the Central Museum of the People's Army of Vietnam in Hanoi. In 1992, Schweitzer, using a data scanner provided by the U.S. government, scanned thousands of photographs and supporting documents that have allowed the DPMO to resolve many discrepancy cases (McConnell 1995).

Hoaxes

The families of POWs who were later repatriated and those of still-missing servicemen, as well as investigators of this issue and the general public, have been the targets of opportunists and swindlers. As early as 1976, the House Select Committee on Missing Persons concluded, "MIA families and the American public had [*sic*] been misled too long and too often by charlatans, opportunists, intelligence fabricators and publicity mongers, who preyed on the hopes and sorrows of patriotic citizens" (*Americans Missing* 1976a, 21, 44). Nearly 20 years later, Donnie K. Collins, the wife of a repatriated POW, described before a congressional committee the activities of con artists "who were constantly preying directly or indirectly on the families."

> Their activities range from outright scams to irresponsible statements and alleged information, such as photos or firsthand sightings, which turned out to be false and almost always played a cruel hoax on some or all family members by raising false hopes even if they did not outright steal money. I once caught such a person who had organized dedicated and concerned young people to collect money on the street "for the POWs," which he pocketed. I caught a group stealing and selling POW bracelets for their own profit. I have seen impostors claiming to be returned POWs and sponging off the public. There have been many "POW rescue" missions organized and led usually by kooks and based on false information and usually bilking some families out of money and, again, creating a cruel hoax of false hopes. Also sharing the blame for misinformation leading to false hopes are the many irresponsible statements over the years by public officials, primarily in the Congress, who were either duped themselves by misleading information or were seeking political gain. (*Oversight Hearings* 1994, 797)

During and since the war, information peddlers have approached family members and those investigating the issue with offers to sell information, photographs, or artifacts (Colvin 1987, Keating 1994, McConnell 1995). Determining the validity of any information and its source poses serious challenges for investigators, let alone emotionally involved family members. Take, for example, the live-sighting issue. Investigating reports of live American prisoners being held in Southeast Asia has occupied the DIA POW/MIA Office and its successor, the DPMO, for years. The DPMO has spent a large percentage of its resources investigating more than 1,750 live-sighting reports (*DIA* 1991, 219). The agency determined that a significant number of live-sighting reports were inconclusive, and others were the work of a foreign intelligence agency that wanted to prevent improved relations between the United States and Vietnam (Franklin 1992, 119).

Moreover, according to the Senate Select Committee, "the manufacture of POW/MIA related materials, including photographs, dog tags and other purported evidence of live Americans, has become a cottage industry in certain parts of Southeast Asia, particularly Thailand" (Senate Select Committee 1993, 34). During the 1980s, the POW/MIA Office received over 5,000 communications from persons, many of them living in Vietnam, who contended that they possessed dog tags or other forms of military identification from MIAs. Upon investigation, only 6 percent of these reports pertained to soldiers KIA/BNR, and 4 percent to MIAs. Most dog tags pertained to servicemen who returned alive, were killed in action, or were repatriated POWs. The DIA contended, "Years of investigation and analysis have shown that dog tag reports have been instigated by elements of Vietnam's government in an effort to influence and exploit the POW/MIA issue" (U.S. Department of Defense, *Factbook*, 1990, 21).

In 1992, the national directorship of the Veterans of Foreign Wars of the United States (VFW) indicated that individuals falsely claiming to be former POWs have approached VFW posts, soliciting funds "to free an American POW from a Vietnam jail" (*Oversight Hearings* 1994, 351). McConnell (1995, 182) also cites instances of fake POWs using their alleged status to further their own ends. The POW/MIA organizations themselves are not immune from criticism, either. At one time, 26 percent of the people on the membership rolls of the Arlington, Texas, based American Ex-POW Association were fakes (McConnell 1995, 182). During its investigation of the fund-raising activities of POW/MIA groups, the Senate Select Committee found that "in some instances an excessive percentage of funds was retained by the fundraising organization. In others, the fundraising solicitations have over-stated to the point of distortion the weight of evidence indicating that live U.S. POWs continue to be held in Southeast Asia" (McConnell 1995, 164– 166; *Senate Select Committee* 1993, 34). Some activities by POW/MIA activists, such as offering large rewards, may have prompted hoaxes and grave robbing, complicating U.S. government efforts to resolve the issue (McConnell 1995, 167).

Congressional Activity

The U.S. Congress has played an important role in relations with Vietnam, and a number of congressmen have made veterans affairs and the POW/MIA issue a topic of particular concern. For example, on 7 August 1964, Congress passed the Tonkin Gulf Resolution, which gave broad authority to the president of the United States—in this case, Presidents Lyndon B. Johnson, John F. Kennedy, and Richard Nixon—to use any level of force to aid South Vietnam and other U.S. allies in Southeast Asia. Congress repealed the Tonkin Gulf Resolution in 1970. Yet as the conflict progressed, Congress sought to limit the president's power, an unprecedented situation for a U.S. commander-in-chief during a time of military action. With the Cooper-Church Amendment to the fiscal year 1971 defense appropriations bill, Congress banned the use of U.S. ground troops in Laos or Thailand. By 1973, Congress had passed ten other laws that set deadlines on U.S. activities in Vietnam, such as the Case-Church Amendment, which barred the appropriation of funds for U.S. military combat in Southeast Asia unless congressional approval was given. Signing of the Paris Peace Accords took place in January 1973, and on 7 November of that year Congress passed, over the veto of President Nixon, the War Powers Resolution, which required that both Congress and the president collectively determine the necessity and level of U.S. military activities.

Congressional approval has also been required for the numerous hearings that have taken place on the POW/MIA issue since the 1970s. Both houses of Congress, but particularly the House of Representatives, have taken an oversight role vis-à-vis the POW/MIA issue. In the House of Representatives, the House Select Committee on Missing Persons in Southeast Asia—known as the Montgomery Committee—and the House Subcommittee on Asian and Pacific Affairs conducted hearings on such issues as POW/MIA affairs, Cambodian refugees, and human rights in Southeast Asia. From the 1970s to the 1990s, the subcommittee sponsored the POW/MIA Task Force that held more than 50 hearings at which the testimony of over 100 people was heard. These witnesses included former prisoners, veterans, government representatives, the leaders of nongovernmental agencies active in Southeast Asia, the National League of Families of American Prisoners and Missing in Southeast Asia, and Southeast Asian Refugees.

During the 1975–1977 period, the Montgomery Committee,

chaired by Gillespie V. Montgomery (D-MS), attempted to determine if POW or MIA remains were being held by the Vietnamese government. It conducted a wide range of hearings; met with North Vietnamese, Cambodian, and Laotian officials; and examined primary intelligence sources. While some members dissented from the majority decision, the committee in its final report concluded that "no Americans are still being held alive as prisoners in Indochina, or elsewhere, as a result of the war in Indochina" (*Americans Missing* 1976a, 238).

In addition, the House Armed Services Committee held hearings on the problems of POW/MIA families (1970) and the activities of the Central Identification Laboratory (1986–1987). The House Committee on International Relations Subcommittee on Asian and Pacific Affairs likewise held hearings on policies and procedures (1979–1980), fact-finding missions (1986–1987), access to classified live-sighting information (1988), operations on the Defense Intelligence Agency (1991), Soviet knowledge of Korean War and Vietnam conflict POWs (1992–1993), and future relations with Vietnam (1994). By 1995, oversight of activity on the POW/MIA issue became the responsibility of the House Committee on National Security Military Personnel Subcommittee, chaired by Robert Dornan (R-CA). This committee heard testimony by POW/MIA family members about individual cases, the activities of the DPMO to resolve outstanding cases, the level of cooperation by the Vietnamese to resolve cases, and the possibility that POWs from the Korean War and Vietnam conflict were sent to the Soviet Union (1995–1996).

The Senate held fewer and more sporadic hearings but was a primary venue for debates about normalization. In 1985, the Senate Foreign Relations Committee focused on live-sighting reports and an alleged U. S. government conspiracy and cover-up. The committee reached no definite conclusions. The Senate Select Committee on POW/MIA Affairs (1991–1992), co-chaired by Senators John Kerry (D-MA) and Robert Smith (R-NH), likewise focused on U. S. government policy and practices, the question of possible American prisoners being held in Southeast Asia after the war's end, declassification of relevant documents, the adequacy of the POW/MIA accounting process, and the activities of private organizations on this issue. In its conclusions, the committee could not rule out the possibility that live Americans are being held in Southeast Asia. It also found that the U. S. government agencies responsible for accounting for POW/MIAs had been variously effective or ineffective depending on the priorities and resources allocated by the presidential administration in power at the time. The committee supported further archival research and joint field activities with representatives of the governments of Vietnam, Laos, and Cambodia.

During the 1990s, Congress had its say about the POW/MIA issue. In 1991, it passed the McCain Amendment to the annual Defense Authorization Act, which required the Department of Defense to declassify any record, live-sighting report, or other information relating to the location, treatment, or condition of any Vietnam-era POW/MIA. In January of 1994, Congress debated the future of the trade embargo against Vietnam and the McCain Amendment, a nonbinding resolution to the Foreign Relations Authorization Act, which expressed the sense of the Senate that "in order to maintain and expand further United States and Vietnamese efforts to obtain the fullest possible accounting of American servicemen unaccounted for during the war in Vietnam, the President should lift the United States trade embargo against Vietnam immediately" ("Foreign Relations Authorization Act" 1994, S211). The next week President Clinton announced the lifting of the trade embargo. During 1996 Congress overhauled procedures that the Department of Defense uses to determine the status of missing service members, DOD civilian employees, and DOD contractor personnel (see Chapter 4). The following year the Senate confirmed the nomination of Douglas "Pete" Peterson as the first U.S. ambassador to the Socialist Republic of Vietnam.

Presidential Perspectives

Securing the release of American prisoners ranked high on President Nixon's list of priorities in negotiating the peace in Vietnam, and the return of prisoners during Operation Homecoming was the fruit of lengthy, painstaking negotiations by Henry Kissinger. Prior to being forced to resign by the Watergate scandal in 1974, President Nixon told the American people that all American prisoners had been returned; but in a cable to the North Vietnamese Prime Minister Pham Van Dong, he voiced his incredulity that only 10 of the repatriated POWs were from Laos when U.S. intelligence sources indicated that approximately 300 Americans had been captured and may have been held in Laos (Senate Select Committee 1993, 84).

During President Gerald Ford's administration (1974–1976), efforts to determine the fate of MIAs were hampered by a lack of resources. "The withdrawal of U.S. forces from Vietnam in 1973 and the fall of Saigon in April 1975 resulted in a vast reduction in the level of field assets and the opportunity for access to geographic locations as well as human resources" (U.S. Department of Defense, *Factbook*, 1990, 19). In addition, Congress forbade the use of U.S. forces in Southeast Asia (*Americans Missing* 1976a, 149). As stipulated by the Paris Peace Accords, in 1974 and 1975 the government of North Vietnam unilaterally repatriated 26 sets of identified remains of American prisoners who had died in captivity; yet, it resisted U.S. efforts to make North Vietnam accountable for American prisoners who had been captured in Laos and Cambodia. When North Vietnam broke the Paris Peace Accords in 1974 by invading and conquering South Vietnam, President Ford imposed on all of Vietnam the same level of sanctions as had been previously applied only to the North. Sanctions included seizing all Vietnamese assets in the United States, rejecting demands for aid, and vetoing requests by the Vietnamese for membership in the United Nations (*Accounting* 1996a, 370).

Taking a less confrontational stance, President Jimmy Carter (1976–1980) sent the Presidential Commission on American Missing and Unaccounted for in Southeast Asia, known as the Woodcock Commission, to press the Vietnamese to account for the missing as a humanitarian concern. After the Woodcock Commission declared that no live Americans existed in Southeast Asia, and the Montgomery Committee split on the issue, the matter appeared to be one of repatriating remains. As then allowed by law, the military status review process changed the status of remaining missing servicemen to KIA/BNR through a presumptive finding of death if no new information could be learned about the missing within a set period of time. The results of this review were later challenged in the courts by some families of MIAs, and the missing were returned to MIA status—to the delight of some families and the chagrin of others (*Americans Missing* 1976a, 182–187). In 1977, President Carter initiated normalization talks, offering advance incentives to the Vietnamese for resolving the MIA issue, such as the possibility of removing travel restrictions, of no longer opposing Vietnam's application for membership in the United Nations, and finally of lifting the trade embargo. During the year the normalization talks took place, the Vietnamese government repatriated 45 identifiable sets of remains (*Accounting* 1996a, 409), and Vietnam was admitted to the United Nations. However, when the Vietnamese government pressed its demands for reconstruction aid, the U.S. government insisted that a full accounting of MIAs take place first. An uproar ensued in Congress in which representatives expressed a vehement refusal to grant aid to Vietnam. After Vietnamese troops invaded Cambodia in 1978 and Vietnam aligned itself with the Soviet Union, normalization talks ground to a halt (Stern 1995, 25), and loans to Vietnam by the International Monetary Fund and World Bank ceased (McConnell 1995, 84). No further remains were returned during the Carter administration.

President Ronald Reagan (1981–1988) gave new impetus to resolving the POW/MIA issue by denoting "the fullest possible accounting of our missing men a matter of the highest national priority" (U.S. Department of Defense, *Factbook,* 1990, 11). When Robert Garwood, a U.S. Marine private who had remained in Vietnam and was later tried and convicted of collaborating with the enemy, returned to the United States in 1979, the event made people reconsider the possibility of Americans being held captive in Southeast Asia. Moreover, in 1981, the congressional testimony of General Eugene Tighe, the retiring head of the Defense Intelligence Agency who personally believed that intelligence data indicated that Americans were possibly being held captive in Southeast Asia, further eroded public confidence in the U.S. government's statements to the contrary.

President Reagan's desire to resolve the POW/MIA issue was evident in the increase in resources he allocated to it. He tripled the number of personnel involved in the Joint Casualty Resolution Center (JCRC), the DIA POW/MIA Office, and the Central Identification Laboratory in Hawaii (CILHI). He included a representative of the National League of Families of American Prisoners and Missing in Southeast Asia in the Interagency Group that set policy on this issue, consulted regularly with Congress, and asked for diplomatic third-party efforts on behalf of the United States to persuade the Vietnamese to cooperate. Although Reagan made the American public aware of the issue and his efforts to address it by renewing negotiations with the Vietnamese, he eschewed private efforts to rescue alleged prisoners through force. Dialogue with the Vietnamese in the early 1980s focused on treating the MIA issue as a humanitarian one and putting into effect the research and excavation activity the two countries agreed to in principle (Stern 1995). By 1984, the Vietnamese government had unilaterally repatriated another 21 identifiable sets of remains, agreed to joint investigations of crash sites, and acquiesced to U.S. demands to release Vietnamese citizens who had been interned in "reeducation" camps for opposing communism and allow them to emigrate to the United States.

In 1987, with the unanimous consent of Congress, Reagan appointed General John Vessey as a special emissary to Vietnam (see Chapter 3). The Vietnamese government agreed to investigate discrepancy cases and between 1985 and 1988 repatriated another 122 sets of identifiable remains, one of which was the result of a joint field excavation (*Accounting* 1996a, 411). At this juncture, the Vietnamese government again attempted to link POW/MIA accountability to the U.S. government's direct providing of humanitarian assistance. (Although Congress had prohibited direct governmental assistance to Vietnam, it allowed nongovernmental agencies to provide humanitarian assistance in the form of child survival and prosthetics projects [Stern 1995, 41–43].) In its *Final Interagency Report,* the Reagan administration indicated that resolving discrepancy cases had become a priority of both the United States and Vietnam. Yet the Reagan administration took the position that "because of such discrepancies and the lack of knowledge about many cases, the Reagan Administration has concluded that we must operate under the assumption that at least some of the missing could have survived until we can jointly conclude that all possible efforts have been made to resolve their fate" (21).

During the administration of President George Bush (1989–1992), the premise remained the same as it had under the Reagan administration: The U.S. government would not discount the possibility that "some Americans may be alive in Indochina," though "no single report or combination of reports and technical sensors has thus far been specific enough to offer conclusive proof that Americans remain in captivity" (U.S. Department of Defense, *Factbook,* 1990, 20). It appeared for the first time that the Vietnamese were truly planning to withdraw their troops from Cambodia, which was a precondition for improved relations with the United States, and the Vietnamese leadership involved in the POW/MIA issue changed favorably, allowing American access to the central Vietnamese military archives. The Bush administration further expanded resources and personnel to deal with accounting for the missing, creating in 1991 the Joint Task Force–Full Accounting (JTF-FA) with task force detachments in Hanoi; Bangkok, Thailand; Vientiane, Laos; and Phnom Penh, Cambodia (Stern 1995, 73).

The Bush administration then formulated what became known as the "road map," which set out steps for Vietnam to take and the reciprocal steps the United States would make to normalize relations between the two countries. However, the Vietnamese did not agree with what they called the unilateral nature of the road map. The road map required full cooperation by the Vietnamese in resolving discrepancy cases and repatriating all recoverable remains. In response to such cooperation, the United States would lift a 25-mile ban on travel by Vietnamese Foreign Ministry officials from their United Nations Mission in New York City and relax restrictions on travel by groups to Vietnam. Despite a lack of full accounting, the United States took these steps in 1991 after a peace agreement was reached in Cambodia (*Accounting* 1996a, 380–381). It also allowed further humanitarian relief through nongovernmental agencies and medical care projects by the military during field investigations and made exceptions to the trade embargo for telecommunications, basic necessities, and the activities of nongovernmental humanitarian organizations (Stern 1995, 79, 81). In December 1992, the White House announced that it would allow businesses to open offices in Vietnam to conduct preliminary activities, such as negotiating contracts that could be finalized after the eventual complete lifting of the trade embargo (Stern 1995, 91).

Nevertheless, Vietnam and the United States continued to take different approaches to the question of accountability. The Vietnamese considered a case resolved if there existed some evidence of a person's death, short of having the actual remains, but the United States demanded a chain of evidence to show that an individual had died, and then went on to consider if remains were recoverable (Stern 1995, 57). During the Bush administration, 65 sets of identifiable remains were repatriated, 12 of which were the result of joint field activities (*Accounting* 1996a, 45).

Today's Debate on Vietnam MIAs

Arguments for and against the existence of prisoners held after the signing of the Paris Peace Accords revolve around a number of issues. Among these are the previous experiences of captives held by Communist forces, interpretations of intelligence data, debate about prisoners being sent to other countries, and the issue of deserters and collaborators. Whether or not any such prisoners could have survived to the present day is another topic of debate.

After the Indochina War (1946–1954), hundreds of French soldiers from France, its colonies, and the French Foreign Legion stayed in Vietnam as deserters and collaborators, marrying Vietnamese women. Others, such as Foreign Legionnaires from Eastern Europe and Germany, were repatriated to their country of origin rather than to France, which makes it appear as if more personnel remained in Vietnam than truly did (Franklin 1992, 25–26). While Sauter and Sanders (1993, 27–28) assert that the French government later learned that intelligence officers and technicians had been held against their will, a U.S. government report maintained that those who remained in Vietnam did so voluntarily (*Accounting* 1996a, 531). It is documented that the North Vietnamese government charged the French government exorbitant sums for the maintenance of French military cemeteries and repatriated remains depending on the French government's response to the North's political activity (Clarke 1979, 55–58; *Accounting* 1996a, 535–536). A factor in the repatriations of remains to France was the fact that North Vietnam was at war with the South and its supporters during the post–Indochina War period, and the transportation of remains to France would have been difficult at best (Franklin 1992, 27).

Some investigators who believe that American prisoners from World War II, the Korean War, and the Cold War were sent to the Soviet Union to be pumped for intelligence information and to be used as guinea pigs in brainwashing experiments also maintain that during the Vietnam conflict POWs were sent to the Soviet Union (Brown 1993; Jensen-Stevenson and Stevenson 1990; Sanders, Sauter, and Kirkwood 1992; Sauter and Sanders 1993) and that a separate camp system was used for this purpose (Sanders, Sauter, and Kirkwood 1992). The authors of a report by the joint American-Russian Task Force Russia, released by the Pentagon in 1993, concluded, "The Soviets transferred several hundred U.S. Korean War POWs to the U.S.S.R. and did not repatriate them. This transfer was mainly politically motivated with the intent of holding them as political hostages, subjects for intelligence exploitation, and skilled labor within the camp system" (Department of Defense Prisoner of War/Missing in Action Office 1993, 39). Based on the work of Task Force Russia, the Senate Select Committee found that "there is strong evidence, both from archived U.S. intelligence reports and from recent interviews in Russia, that Soviet military and intelligence officials were involved in the interrogation of American POWs during the Korean Conflict. . . . Additionally, the Committee has reviewed strong evidence that some unaccounted for American POWs from the Korean Conflict were transferred to the former Soviet Union in the early 1950s." The committee stated further that some Korean War and Cold War prisoners might still be alive in the Soviet Union (Senate Select Committee 1993, 37).

In regard to the Vietnam conflict, it has been confirmed that the Soviets had a larger military role in Vietnam than previously had been believed. At least one American prisoner of war was directly interrogated by a Soviet intelligence officer, and others were thought to have participated indirectly in prisoner interrogations (McConnell 1995, 282–285; Senate Select Committee 1993, 37). Former National Security Agency analyst Jerry Mooney has frequently asserted his belief that prisoners might have been transferred to the Soviet Union during the war, an assertion that has often been cited by activists but has not been proven through Mooney's data (McConnell 1995, 267). Furthermore, Jan Senja, a Czechoslovakian major general who defected to the West in 1968, has also alleged that up to 90 American prisoners of war were sent to the Soviet Union via Prague (McConnell 1995, 286–287; Senate Select Committee 1993, 426). Russian President Boris Yeltsin, in a letter to the joint Russian-American commission, stated that the Russian government had found evidence of Americans being held prisoner from World War II and the Cold War and that several Americans taken captive during the Vietnam conflict had also been held in the U.S.S.R. Further efforts to investigate the Russian connection were stymied by Russian officials (McConnell 1995, 281).

Several U.S. government employees working for the military and intelligence agencies expressed their opinion that POWs were left behind and that this information is supported by intelligence data. They include General Eugene Tighe, Garnett "Bill" Bell, and Jerry Mooney, whose assertions are contained in the documents of the Senate Select Committee on POW/MIAs and in many works by activists. Moreover, a Vietnamese mortician who defected to the United States when he was being expelled from Vietnam due to his Chinese ancestry reported on prisoners and passed lie-detector tests (*Accounting* 1996a, 496). Not only as an employee of the Directorate of Cemeteries for Hanoi had he processed several hundred sets of remains of American servicemen, he reported having seen three men whom he was told had desired to remain in Vietnam (Sauter and Sanders 1993, 24). According to Sauter and Sanders, one of these men was later determined to be Robert Garwood. At any rate, other sources have collaborated the likely number of remains in the possession of the SRV, a contentious fact that has long stymied efforts by the Vietnamese to deny the existence of remains. The DPMO and its predecessor, the DIA, have over the years investigated aerial photographs, alleged POW photos, and live-sighting reports. While many of these have proven to be inconclusive, or even forgeries, others have merited significant investigative efforts.

The U.S. government has acknowledged the high number of deserters during the Vietnam conflict. Citing a Pentagon report, Sauter and Sanders (1993) described how Americans who had renounced their citizenship were sent to reeducation camps for a three-year period of indoctrination, but they also questioned whether these "students" were truly collaborators or more likely prisoners because they used collaboration as a means of saving their lives, as several prisoners from jungle camps who later returned to the United States during Operation Homecoming might have done. They suggest that this was also the case with Robert Garwood, who upon his return to the United States was convicted of collaborating with the Vietnamese but not of actually deserting to them, and others who chose to stay behind may have done so out of fear of being court-martialed upon their return to the United States (26–28).

The fate of prisoners captured in Laos has remained a deep mystery. Less that 2 percent of the 499 MIAs lost in Laos were returned during Operation Homecoming. This low percentage has led to much speculation regarding their fate: Were all American prisoners captured in Laos sent to North Vietnam or only those along the Ho Chi Minh Trail, an area controlled by North Vietnamese forces? Did the Pathet Lao withhold some prisoners to negotiate separately with the United States? In describing his harrowing ordeal in a Laotian POW camp and subsequent miraculous escape through the three-canopy jungle, Dieter Dengler well demonstrated the difficulties any American prisoner would have encountered (Dengler 1979). Based on his research and personal experience as a prisoner of war of the Vietnamese captured in Laos, Theodore Guy maintained that prisoners under the control of North Vietnamese forces were sorted by location of capture and that the small number of prisoners captured in Laos were a token ("or a huge mistake") because the North Vietnamese did not want to admit to having such a large military presence in Laos (*Accounting* 1996a, 20; see Chapter 4).

In a 1975 speech, Russian dissident Aleksandr Solzhenitsyn expressed his opinion that American prisoners were held after Operation Homecoming:

> If the government of North Vietnam has difficulty explaining to you what happened to your brothers, your

American POWs who have not yet returned, I can explain this quite clearly on the basis of my own experience in the Gulag Archipelago [the Soviet prison system]. There is a law in the Archipelago that those who have been treated the most harshly and who have withstood the most bravely—who are the most honest, the most courageous, the most unbending—never again come out into the world. They are never again shown to the world because they will tell tales that the human mind can barely accept. Some of your returned POWs told you that they had been tortured. This means that those who have remained were tortured even more, but did not yield even an inch. These are your best people. These are your foremost heroes who, in a solitary combat, have stood the test. (cited in Boettcher and Rehyansky 1981, 958)

When the SRV allowed those in charge of the military archives in Hanoi to funnel documents through a private channel to the United States, it held back certain documents that were listed in the main index of the archive. Retained were files on individual prisoners that hold records—and audiotapes—of interrogations. This tactic and anecdotal evidence suggest that a large number of prisoners were executed, tortured to death, or ripped apart by mobs (McConnell 1995). If this is so, SRV reluctance for the truth to be known is understandable because the communist government would be condemned in the court of world opinion, and perhaps other courts as well.

Although U.S. government sources hold the position that the likelihood of any prisoners surviving to this day is dim, the official position has long been not to discount the possibility (Senate Select Committee 1993, 7). Retired Air Force officer and former prisoner of war in Vietnam, Theodore Guy, citing the 20-year captivity and survival of South Vietnamese commandos in communist "reeducation" camps (*Accounting* 1996a, 9), asserted that he believes American military personnel could also survive such horrible conditions with proper morale (*Accounting* 1996a, 11–12; see Chapter 4). Many POW/MIA veterans and activist organizations hold this position (see Chapter 5). Their slogan remains, "Bring 'em Home."

Political Aftermath

When President Bill Clinton took office in January 1993, he inherited the Bush "road map." President Clinton indicated that the new administration would follow the road map policy of preconditions to discussions of normalization. In March 1993, Secretary of State Warren Christopher remarked that Vietnam had met the first of the preconditions, that is, the withdrawal of Vietnamese forces from Cambodia so a political settlement could be made in that country. United Nations–sponsored elections took place in Cambodia in 1993.

The U.S. government had long blocked lending to Vietnam from the International Monetary Fund (IMF), a policy supported by the Trading with the Enemy Act, which had deterred the Vietnamese government from getting the loans it needed to improve the country's infrastructure. Yet by 1993, many countries were no longer observing the trade embargo and were willing to lend large sums to Vietnam through these international lending institutions. Thus, the ability of the United States to limit lending to Vietnam had diminished over the years. When on 2 July the United States discontinued its objections to lending to Vietnam, the IMF, World Bank, and Asian Development Bank were poised for action. According to F. Gale Connor, "Once Vietnam again became eligible for loans from international lending institutions, to which the United States is a major contributor, an overhaul, if not outright abandonment, of the embargo became inevitable. Otherwise, U.S. tax money would have been committed to development projects in Vietnam for which U.S. companies would have been unable to bid" (Connor 1994, 486).

In mid-July, a U.S. presidential delegation led by Deputy Secretary for Veterans Affairs Hershel Gober and Assistant Secretary of State Winston Lord met with the Vietnamese general secretary, minister of defense, minister of the interior, and deputy foreign minister. After the U.S. representatives reiterated the administration's policy on remains, discrepancy cases, trilateral cooperation with Laos, and archival research, the Vietnamese pledged cooperation but professed that they expected limited results (Stern 1995, 109).

In September 1993, President Clinton renewed the trade embargo against Vietnam, citing a lack of progress in the POW/MIA area. An exception was made for U.S. corporations that would be bidding on projects funded by international financial institutions.

Yet by the New Year, the picture had changed. On 27 January

1994, the U.S. Senate passed a "sense of the Senate" resolution that urged President Clinton to lift the trade embargo against Vietnam (see Winston Lord's Speech, Chapter 4). The following week, on 3 February 1994, President Clinton lifted the embargo. "In view of this tangible progress in all four of the identified areas and the development of mechanisms capable of achieving that fullest possible accounting, the President judged that the best way to ensure continued cooperation and encourage even greater progress was to lift the embargo and open the door to the establishment of a U.S. Liaison Office in Hanoi," testified Kent Wiedemann, deputy assistant secretary of state, Bureau of East Asian and Pacific Affairs (*Accounting* 1996a, 193).

In 1994 and 1995, the U.S. government continued to cite "excellent" cooperation by the Vietnamese, and progress with the Laotians and Cambodians as well. Then DPMO Director James Wold described archival research activities, joint excavations, the amnesty program for citizens to turn in remains to the Vietnamese government, and the investigations of live-sighting reports (*Accounting* 1996a, 212–228). During the first Clinton administration, 121 sets of identifiable remains were repatriated, although some family members disputed the validity of the identifications by CILHI (see Cressman testimony, Chapter 4). On 11 July 1995, when President Clinton announced that the United States would normalize relations with Vietnam (see Chapter 4), he also indicated that only 55 discrepancy cases remained unsolved.

Supporters of normalizing relations with the Socialist Republic of Vietnam advanced several arguments, including improved cooperation on the POW/MIA issue, economic and strategic interests in the region, and human rights. In 1994, Theodore Schweitzer, the private citizen through whom the Vietnamese had passed documents from the Central Military Museum in Hanoi to the United States, testified before a congressional panel that many sets of MIA remains and relevant artifacts and documents were in private hands and that "the more steps the United States takes to ease the hardships on Vietnam, the more warmth the common Vietnamese citizens will feel towards us and will come forward with materials" (*Oversight Hearings* 1994, 1186). Since the lifting of the trade embargo, the U.S. government has called the cooperation of Vietnam on the POW/MIA issue good. "Just two months ago [May 1995], officials from the Departments of State, Defense, and Veterans Affairs traveled to Asia for high-level talks with their counterparts in both Vietnam and Laos," Senator Tom Daschle (D-SD) told the Senate on 12 July 1995. "During that trip they were presented with more than 100 documents, which the Defense Department has called the most detailed and informative turned over to date. Moreover, our officials characterize the cooperation they had received from the Vietnamese as excellent" (Daschle 1995, S9727). About the efforts of those in the DPMO, Senator John McCain (R-AZ) told Congress on 11 July 1995, "It is mostly my faith in the service of these good men and women that has convinced me that Vietnam's cooperation warrants the normalization of our relations under the terms of the roadmap" (McCain 1995, S9723). McCain also stated that the U.S. government's ability to affect strategic concerns and human relations concerns would be enhanced by restoring relations. Noting that the United States would be the 161st nation to recognize Vietnam, Alan Simpson (R-WY) remarked on the economic advantages already accrued to those other countries and commended the Veterans of Foreign Wars for supporting normalization. "We are of the opinion if normalizing relations with Vietnam furthers the process toward fullest possible accounting, then we would support this decision," the organization maintained in a statement released on 13 June 1995 (Simpson 1995, S9731). "Recognizing Vietnam does not mean forgetting our MIAs, by any means. Recognizing Vietnam does not mean that we agree with the policies of the Government of Vietnam. But recognizing Vietnam does help us promote basic American values, such as freedom, democracy, human rights, and the marketplace," Frank Murkowski (R-AK) told the Senate on 11 July 1995. "When Americans go abroad or export their products, we export an idea, a philosophy, and a government. We export the very ideals that Americans went to fight for in Vietnam" (Murkowski 1995, S9649).

On the other hand, many people have been critical of President Clinton's decision to normalize relations with Vietnam. Their reasons include what they see as a resultant lack of leverage to extract information from the Vietnamese and the reality that the Vietnamese are holding back remains and documents that could unilaterally resolve discrepancy cases (see Chapter 4, Griffiths). Longtime champion of MIA families, Senator Robert Smith declared normalization of relations a course of action "nothing short of capitulation to Communist blackmail."

> I do not see what incentive Vietnam will have to disclose additional, perhaps damaging, information on

POW/MIAs if they get what they want from us first. In short, I believe that the answers to the POW/MIA issue will only be obtained through serious negotiations with Vietnam after it is made clear to Hanoi's Communist Politburo that we are not going to go any further to meet their political agenda without first getting full disclosure on our POWs and MIAs. We should also be taking a similar approach with the Government of Laos. (*Accounting* 1996a, 139)

"An incentive first approach has never worked with Vietnam," wrote Richard Childress, director of Asian Affairs, National Security Council during the Reagan years, "only one of strict reciprocity, the basis of the roadmap developed in the Bush administration" (*Accounting* 1996a, 865).

During a congressional hearing in June 1996 researchers Garnett "Bill" Bell and George J. Veith summarized the prevailing arguments that Vietnam is not cooperating.

Although the U.S. government claims that Vietnam is doing everything it can to account for the 2,200 American personnel still unaccounted for in Indochina, this contention is not supported by facts. On the contrary, all available evidence suggests that the Vietnam Communist Party could rapidly account for a significant number of MIA cases, especially the 95 men associated with the "Special Remains" cases, who either died due to disease or were executed in wartime prison camps, or whose remains have been depicted in photographs released by Vietnam. Evidence of a complex wartime record keeping system indicates that Vietnam could also provide important information on many of the 305 last-known-alive discrepancy cases, as well as crash and grave sites. (*"Full Faith"* 1997, 210–211)

John F. Sommer Jr., executive director of the American Legion, speaking on behalf of the organization's nearly 3 million members, accused the Vietnamese of using the MIA issue for monetary gain:

The Administration continues to praise the "cooperation" the Vietnamese are providing the Joint Task

Force–Full Accounting in helping resolve the POW/MIA issue. The cooperation Vietnam is providing is that for which they are being paid substantial sums of money by the United States. This consists of joint U.S.-Vietnamese excavation of crash sites. The cooperation that is needed . . . is unilateral Vietnamese recovery and repatriation of remains, and unilateral actions on collecting and turning over documents, particularly those which would be helpful in resolving cases of missing American Servicemen. (*Accounting* 1996a, 265)

Indeed, the cost of joint excavations is staggering. One source estimated the cost at $1.7 million per MIA case solved (McConnell 1995, 215). With joint excavations proving to be such a lucrative enterprise, little incentive exists for the Vietnamese to turn over remains unilaterally, criticized Ann Mills Griffiths (see Chapter 4).

Looking Ahead

It is a product of our human nature to want answers, to discern the truth of events that have affected us in emotional, often traumatic, ways. As a species we attempt to improve our ability to survive by learning from our history. There is much yet to be learned from the Vietnam war experience in general and from the POW/MIA issue in particular.

The ability of citizens to probe the military aspects of the war has been limited by the U.S. government's need to maintain the security of its intelligence operations and perhaps too by its desire to hide its failures. Nevertheless, over more than two decades families of MIAs and other interested citizens have struggled to get satisfactory answers. In so striving they have brought to the attention of the larger public questions about the relationship of the U.S. government to its military forces—questions of loyalty and commitment. Whether families of MIAs will ever be truly satisfied with the answers they find is doubtful, but that they will continue to search for these answers is a fact. Even as the list of Vietnam-era MIAs slowly dwindles, some families of the more than 8,000 MIAs from the Korean War and the Cold War have begun to investigate the fates of their loved ones. Thus, no clear end to the MIA question is in sight.

References

Alvarez, Everett Jr., and Anthony S. Pitch. 1989. *Chained Eagle*. New York: D. I. Fine.

Anton, Frank, with Tommy Denton. 1997. *Why Didn't You Get Me Out? Betrayal in the Viet Cong Death Camps*. Arlington, TX: Summit Publishing Group.

Barker, A. J. 1974. *Behind Barbed Wire*. London: B. T. Batsford.

Blakey, Scott. 1978. *Prisoner of War: The Survival of Commander Richard A. Stratton*. Garden City, NY: Anchor Press, Doubleday.

Boettcher, Thomas D., and Joseph A. Rehyansky. 1981. "We Can Keep You . . . Forever." *National Review* (21 August): 958–962.

Brace, Ernest C. 1988. *A Code to Keep*. New York: St. Martin's Press.

Brown, John M. G. 1993. *Moscow Bound*. Eureka, CA: Veterans Press.

Cawthorne, Nigel. 1991. *The Bamboo Cage: The True Story of American P.O.W.s in Vietnam*. New York: S.P.I. Books.

Chesley, Larry. 1973. *Seven Years in Hanoi: A POW Tells His Story*. Salt Lake City, UT: Bookcraft.

Clarke, Douglas L. 1979. *The Missing Man: Politics and the MIA*. Washington, DC: National Defense University.

Colvin, Rod. 1987. *First Heroes: The POWs Left Behind in Vietnam*. New York: Irvington Publishers.

Connor, F. Gale. 1994. "Vietnam: Trading with the Enemy or Investing in the Future?" *Law and International Business*, no. 2 (winter): 481–489.

Daschle, Tom. 1995. "United States-Vietnam Relations: Looking Forward." *Congressional Record* 141, no. 112 (12 July): S9727–S9728.

Dengler, Dieter. 1979. *Escape from Laos*. San Rafael, CA: Presidio Press.

Department of Defense Prisoner of War/Missing in Action Office. 1993. Joint Commission Support Branch, Research and Analysis Division, "The Transfer of U.S. Korean War POWs to the Soviet Union," 26 August.

Final Interagency Report of the Reagan Administration on the POW/MIA Issue in Southeast Asia. 1989. Washington, D.C.

"Foreign Relations Authorization Act." 1994. *Congressional Record* 140, no. 3 (27 January): S211.

Franklin, H. Bruce. 1992. *M.I.A. or Mythmaking in America*. New York: Lawrence Hill Books.

Jensen-Stevenson, Monika, and William Stevenson. 1990. *Kiss the Boys Goodbye: How the United States Betrayed Its Own POWs in Vietnam*. Toronto: McClelland and Stewart.

Johnson, Sam, and Jan Winebrenner. 1992. *Captive Warriors: A Vietnam POW's Story*. College Station: Texas A&M University Press.

Keating, Susan Katz. 1994. *Prisoners of Hope: Exploiting the POW/MIA Myth in America*. New York: Random House.

Kern, Timothy. 1989. "Follow-Up: Fifteen Years after Captivity in Southeast Asia." *Journal of the American Medical Association. 19* May: 2776–2784.

Kissinger, Henry. 1982. *Years of Upheaval*. Boston: Little, Brown.

Lippman, Thomas W. 1992. "POW Policy Sowed Lasting Doubts." *Washington Post* (24 September): A6–A7.

McCain, John. 1995. "Restoration of Diplomatic Relations with Vietnam." *Congressional Record* 141, no. 111 (11 July): S9723– S9724.

McConnell, Malcolm, with research by Theodore Schweitzer III. 1995. *Inside Hanoi's Secret Archives: Solving the MIA Mystery*. New York: Simon and Schuster.

McCubbin, Hamilton I., et al., eds. 1974. *Family Separation and Reunion: Families of Prisoners of War and Servicemen Missing in Action*. Washington, DC: U.S. Government Printing Office.

McGrath, John M. 1975. *Prisoner of War: Six Years in Hanoi*. Annapolis, MD: Naval Institute Press.

Mowery, E., D. Hutchings, and B. Rowland. 1968. "The Historical Management of POWs: A Synopsis of the 1963 U.S. Army Provost Marshal General's Study Entitled 'A Review of the United States Policy on Treatment of Prisoners of War,'" paper. San Diego: Naval Health Research Center.

Mulligan, Jim. 1981. *The Hanoi Commitment*. Virginia Beach, VA: RIF Marketing.

Murkowski, Frank. 1995. "Normalization of Relations with Vietnam." *Congressional Record* 141, no. 111 (20 July): S9648–S9649.

Norman, Geoffrey. 1990. *Bouncing Back*. Boston: Houghton Mifflin.

Risner, Robinson. 1973. *The Passing of the Night: My Seven Years as a Prisoner of the North Vietnamese*. New York: Random House.

Rowan, Stephan A. 1973. *They Wouldn't Let Us Die: Prisoners of War Tell Their Story*. Middle Village, NY: Jonathan David Publishers.

Sanders, James D., Mark A. Sauter, and R. Cork Kirkwood. 1992. *Soldiers of Misfortune*. Washington, DC: National Press Books.

Sauter, Mark, and Jim Sanders. 1993. *The Men We Left Behind: Henry Kissinger, the Politics of Deceit and the Tragic Fate of POWs after the Vietnam War*. Washington, DC: National Press Books.

Simpson, Alan. 1995. "Vietnam and Diplomatic Relations." *Congressional Record*, Vol. 141, no. 112 (12 July): S9730–S9731.

Stern, Lewis M. 1995. *Imprisoned or Missing in Vietnam: Policies of the Vietnamese Government Concerning Captured and Unaccounted for United States Soldiers, 1969–1994.* Jefferson, NC: McFarland.

Stockdale, James. B. 1984. *A Vietnam Experience: Ten Years of Reflection.* Stanford, CA: Hoover Institution.

U.S. Department of Defense. 1990. *POW-MIA Factbook.* Washington, DC: U.S. Government Printing Office.

———. 1955. *POW, The Fight Continues after the Battle, The Report of the Secretary of Defense's Advisory Committee on Prisoners of War.* Washington, D.C.: U.S. Government Printing Office.

———. 1985. *U.S. Casualties in Southeast Asia: Statistics as of April 30, 1985.* Washington Headquarters Services Directorate for Information Operations and Reports. Washington, D.C.: U.S. Government Printing Office.

U.S. House of Representatives. 1996a. *Accounting for U.S. POW/MIAs in Southeast Asia.* Hearing before the Military Personnel Subcommittee, Committee on National Security, 104th Congress, first session, 29 June 1995. Washington, DC: U.S. Government Printing Office.

———. 1976a. *Americans Missing in Southeast Asia: Final Report of the Select Committee on Missing Persons in Southeast Asia,* 94th Congress, second session, 13 December 1976. Washington, DC: U.S. Government Printing Office.

———. 1991. *Examination of Operations at DIA's Special Office for Prisoners of War and Missing in Action.* Hearing before the Subcommittee on Asian and Pacific Affairs of the Committee on Foreign Affairs, 102d Congress, first session, 30 May 1991. Washington, DC: U.S. Government Printing Office.

———. 1976b. *Hearings before the Select Committee on Missing Persons in Southeast Asia: Part 5. Select Committee on Missing Persons in Southeast Asia,* 94th Congress, second session, July 1976. Washington, DC: U.S. Government Printing Office.

———. 1997a. *The Presidential Determination of "Full Faith" Cooperation by Vietnam on POW/MIA Matters.* Hearing before the Military Personnel Subcommittee of the Committee on National Security, 104th Congress, second session, 19 June 1996. Washington, D.C.: U.S. Government Printing Office.

———. 1997a. *Status of POW/MIA Negotiations with North Korea.* Hearing before the Military Personnel Subcommittee of the Committee on National Security, 104th Congress, second session, 20 June 1996. Washington, DC: U.S. Government Printing Office.

———. 1996b. *United States and Vietnamese Government Knowledge and Accountability for U.S. POW/MIAs.* Hearing before the Military Personnel Subcommittee, Committee on National Security, 104th Congress, first session, 14 November 1995. Washington, DC: U.S. Government Printing Office.

U.S. Senate. Committee on the Judiciary. 1997. *Hearing before the Select Committee on Intelligence,* 104th Congress, second session, 19 June 1996. Washington, DC: U.S. Government Printing Office.

———. 1994. *Oversight Hearings: Department of Defense, POW/MIA Family Issues, and Private Sector Issues.* Hearings before the Select Committee on POW/MIA Affairs, 102d Congress, second session, 1–4 December 1992. Washington, DC: U.S. Government Printing Office.

———. 1993. *POW/MIA's: Report of the Select Committee on POW/MIAs,* 103d Congress, first session. Washington, DC: U.S. Government Printing Office.

———. 1972. *Communist Treatment of Prisoners of War, A Historical Survey.* Committee Print, 92d Congress, second session. Washington D.C.: U.S. Government Printing Office.

———. Committee on Veterans Affairs. 1980. *Study of Former Prisoners of War.* Washington, DC: U.S. Government Printing Office.

Zalin, Grant. 1975. *Survivors.* New York: W. W. Norton.

Chronology 2

1945 February: At the Yalta Conference near the end of World War II, President Franklin D. Roosevelt and Chinese nationalist leader Chiang Kai-shek discuss the future control of Indochina, including Vietnam, Cambodia, and Laos.

9 March: Following a Japanese coup d'état against the French government, Bao Dai, a puppet chief of the French, proclaims himself emperor of Vietnam.

July: At the Potsdam Conference, Vietnam is divided into two zones separated along the seventeenth parallel of latitude by a demilitarized zone.

2 September: The Japanese surrender to the Allies, and Communist leader Ho Chi Minh declares the existence of the Democratic Republic of Vietnam (North Vietnam) with himself as president. Bao Dai remains emperor of South Vietnam.

23 September: A. Peter Dewey, American chief of the Office of Strategic Services (a forerunner of the Central Intelligence Agency) in Saigon (now Ho Chi Minh City), is shot by the Communist Viet Minh soldiers during the conquest of Saigon by French forces, becoming the first American casualty in Vietnam.

1946   February: The Chinese withdraw from North Vietnam and allow the French to return to Tonkin, the region of North Vietnam bordering China.

6 March: The Franco-Vietnamese Accords are signed, making the Democratic Republic of Vietnam a free state within the French Union.

October: Fighting erupts in Haiphong between French and Viet Minh troops.

19 December: The First Indochina War begins when Viet Minh troops attack French forces in Hanoi, the primary city in the French-held Tonkin region of Vietnam.

1947   11 May: Laos becomes an independent state within the French Union.

7 October–22 December: The French attack Viet Minh positions near the Chinese border.

1949   5 June: The French name Bao Dai head of state of Vietnam.

July: The French found the Vietnamese National Army.

October: In China, a civil war pits Mao Zedong's Communist forces against Chiang Kai-shek and his Nationalist Army. Mao Zedong, founder of the Chinese Communist Party, founds the People's Republic of China.

1950   14 January: Ho Chi Minh again declares the establishment of the Democratic Republic of Vietnam. China supplies arms to the Viet Minh.

2 February: U.S. President Truman approves recognition of the Bao Dai government.

May: Truman authorizes $10 million in military aid to the French.

27 June: Truman sends U.S. troops to Korea, beginning U.S. involvement in the Korean War.

3 August: U.S. Military Assistance and Advisory Group arrives in Saigon.

30 December: The United States signs a Mutual Defense Assistance Agreement with France, Vietnam, Cambodia, and Laos.

1951 By this time the U.S. has given approximately $50 million of military assistance to the French in Southeast Asia.

1952 4 November: Dwight D. Eisenhower is elected the U.S. president.

1953 20–26 April: Operation Little Switch to exchange sick and wounded Allied POWs from the Korean War takes place.

27 July: Korean War armistice is signed. Some 8,100 service members are listed as missing.

5 August–6 September: Operation Big Switch repatriates 12,773 Allied POWs.

1954 20–21 July: France signs a cease-fire ending hostilities in Vietnam. Vietnam is divided along the seventeenth parallel of latitude and separated by a demilitarized zone. A ban on arms buildup is put into effect. An International Control Commission of Polish, Canadian, and Indian delegates is chosen; national elections are to be held by 20 July 1956, to choose one government for the entire country.

1955 24 October: Ngo Dinh Diem defeats Bao Dai in a referendum and is proclaimed president of the Republic of Vietnam.

1959   4 April: President Eisenhower first commits to maintaining South Vietnam as a separate nation.

8 July: The first U.S. serviceman is killed by the Vietcong—North Vietnamese Communists—during an attack at Bien Hoa.

31 December: Approximately 760 U.S. military personnel are on assignment in Vietnam.

1960   The Soviet Union provides $200 million in aid to North Vietnam.

20 December: The National Liberation Front (NLF) is created to coordinate efforts to overthrow the Ngo Dinh Diem regime controlling South Vietnam.

1961   21 January: Newly inaugurated President John F. Kennedy approves a Vietnam counterinsurgency plan, which stresses the use of nonconventional warfare tactics.

9–15 May: Vice-President Lyndon Johnson visits South Vietnam and recommends sending more U.S. forces.

15 December: President Kennedy reissues his commitment to a free South Vietnam.

31 December: The official number of U.S. military personnel in Vietnam reaches 3,205.

1962   6 February: A major buildup of American forces begins.

31 December: The official number of U.S. military personnel in Vietnam reaches 11,300.

1963   1 November: A military coup ousts the Ngo Dinh Diem government. Diem is assassinated.

22 November: President Kennedy is assassinated.

31 December: The official number of U.S. military personnel in Vietnam reaches 16,300.

1964   7 August: The Gulf of Tonkin Resolution is passed by the U.S. Congress, allowing President Johnson to use any level of force in aiding South Vietnam and other U.S. allies in Southeast Asia.

31 December: The official number of U.S. military personnel in Vietnam reaches 23,300.

1965   7 February: Vietcong forces concertedly attack U.S. military installations in South Vietnam.

6 April: U.S. ground troops begin offensive operations in South Vietnam.

31 December: U.S. military personnel in Vietnam officially number 184,300; listed as killed in action (KIA) to date are 636 U.S. military service members.

1966   31 December: U.S. military personnel in Vietnam officially number 385,300; listed as KIA to date are 6,644 U.S. military service members.

1967   31 December: U.S. military personnel in Vietnam officially number 485,600; listed as KIA to date are 16,021 U.S. military service members.

1968   31 January: The Tet Offensive launched by the Vietcong inflicts lasting damage to the morale of U.S. and South Vietnamese forces.

16 March: At Mai Lai between 300 and 400 Vietnamese civilians are massacred by U.S. troops.

31 October: The U.S. bombing of North Vietnam is halted.

5 November: Richard M. Nixon wins the U.S. presidential election.

31 December: The official number of U.S. military personnel in Vietnam is 536,600; listed as KIA to date are 30,610 U.S. military service members.

1969 18 March: President Nixon approves the bombing of North Vietnamese forces in Cambodia.

9 April: U.S. troop levels peak at 543,400.

14 May: Nixon proposes peace plan for Vietnam that would leave the country divided.

8 June: Nixon announces the removal of 25,000 U.S. troops from Vietnam, thus beginning the "Vietnamization" of the war.

27 August: The U.S. Ninth Infantry Division leaves Vietnam, beginning Vietnamization of the war.

3 September: Ho Chi Minh dies, leaving the Communist revolution in Vietnam without its famous founder.

31 December: The official number of U.S. military personnel in Vietnam declines to 475,200; listed as KIA to date are 40,024 U.S. military service members.

1970 20 February: Henry Kissinger, President Nixon's advisor on national security affairs, begins secret peace negotiations in Paris.

21 November: A raid by U.S. forces on Son Tay Prison, where POWs were thought to be held in North Vietnam, fails because the prisoners had been moved elsewhere four months earlier.

31 December: The official number of U.S. military personnel in Vietnam falls to 334,600; listed as KIA to date are 44,245 U.S. military service members.

1971 12 November: Nixon limits the use of U.S. ground forces to a defensive role as part of his Vietnamization strategy.

31 December: The official number of U.S. military personnel in Vietnam falls to 156,800; listed as KIA to date are 45,626 U.S. military service members.

1972 23 March: The United States suspends the Paris peace negotiations for lack of results.

27 April: The Paris peace talks resume with renewed vigor.

30 April: U.S. troop levels drop to 69,000.

4 May: The United States suspends peace negotiations and resumes bombing of North Vietnam in retaliation for Vietcong offensives in the central highlands of Vietnam.

13 July: Peace talks resume.

26–27 September: Henry Kissinger conducts secret negotiations with North Vietnamese diplomats in Paris.

19–20 October: Kissinger discusses South Vietnamese support for the Paris Peace Accords with South Vietnamese President Nguyen Van Thieu in Saigon.

7 November: President Nixon wins reelection.

20–21 November: Kissinger and Le Duc Tho refine the Paris Peace Accords.

13 December: The peace talks again halt for lack of progress.

18 December: The Christmas bombing of Hanoi by U.S. aircrewmen begins to bring the North Vietnamese back to the bargaining table.

21 December: The official number of U.S. military personnel in Vietnam falls to 24,000; listed as KIA to date are 45,926 U.S. military service members.

26 December: Peace negotiations resume.

29 December: The bombing of Hanoi ends.

1973   8–12 January: Kissinger and Le Duc Tho negotiate privately, arriving at the Paris Peace Accords that end the war.

15 January: President Nixon halts all U.S. offensive action as a cease-fire goes into effect.

27 January: "The Agreement on Ending the War and Restoring Peace in Viet-Nam" is signed in Paris by representatives of the United States, South Vietnam, North Vietnam, and the National Liberation Front.

12 February: The first American POWs are released by North Vietnam in what is known as Operation Homecoming.

21 February: A peace agreement between Communist Pathet Lao and Royal Lao governments is signed, ending U.S. air strikes in Laos.

29 March: The last American POWs are released by North Vietnam during Operation Homecoming. Last U.S. troops are withdrawn.

1 April: American POWs arrive at Clark Air Force Base in the Philippine Islands for medical treatment.

13 June: The Paris Peace Accords implementation is signed by representatives of the United States, South Vietnam, North Vietnam, and the National Liberation Front.

14 August: All direct U.S. military operations end in Southeast Asia.

31 December: The official number of U.S. military personnel in Vietnam dwindles to 50; listed as KIA to date are 46,163 U.S. military service members.

1974   4 January: South Vietnamese President Thieu asserts that the war in South Vietnam has resumed.

28 January: President Nixon pledges support for the Lon Nol Cambodian government to fight the Communist Khmer Rouge guerrillas.

5–7 April: Khmer Rouge forces take the outposts protecting Phnom Penh, Cambodia. A coalition government is formed in Laos with Pathet Lao in key positions.

3 June: U.S. advisors leave Laos.

9 July: Lon Nol proposes a cease-fire but is rejected by Prince Sihanouk, who is allied with Cambodian Communist leader Pol Pot.

4 August: Nixon resigns from the presidency as a result of the Watergate scandal. Gerald Ford becomes president.

30 November: Lon Nol proposes a cease-fire in Cambodia.

13 December: The North Vietnamese Army (NVA) attacks Phuoc Long, contravening the Paris Peace Accords.

1975 September 1975–November 1976: The Montgomery Committee investigates POW/MIA issue, concluding that no Americans are still being held in Southeast Asia as a result of the war.

21 January: In a press conference, President Ford stresses his unwillingness to reenter the war.

5 March: North Vietnamese troops launch an offensive in the central highlands of South Vietnam.

10 March: The NVA attacks Ban Me Thuot in what proves to be the North's final and successful offensive.

1 April: Lon Nol flees Cambodia for Indonesia, leaving Prince Sihanouk and Pol Pot in control.

9–11 April: Communist insurgents and Laotian troops battle.

10–15 April: Xuan Loc, north of Saigon, is captured by North Vietnamese troops.

1975   12 April: The U.S. ambassador to Cambodia and his
cont.   staff flee their offices there.

17 April: Phnom Penh, Cambodia, falls to the Communists.

20 April: All remaining U.S. military and civilian personnel in Vietnam begin to leave Saigon by means of a military airlift.

21 April: South Vietnamese President Thieu resigns, condemning the United States for abandoning the fight against communism in Southeast Asia.

28 April: Communist leader General Duong Van Minh takes over the government of South Vietnam.

30 April: North Vietnamese troops enter Saigon. President Minh surrenders unconditionally. The Socialist Republic of Vietnam is created.

15 May: U.S. marines rescue the crew of American freighter SS *Mayaguez*, which had been seized by Cambodian Communists.

16 May: Pathet Lao forces seize Pakse in Laos.

June: Pathet Lao forces take over the American embassy in Vientiane, Laos.

23 August: The Communist takeover of Laos is consolidated.

3 December: Laos becomes a Communist state under President Souphanouvong. It is renamed Democratic Kampuchea, but the United States never recognizes it as such.

1977   The Woodcock Commission investigates the POW/MIA issue and concludes that no live POWs exist in Southeast Asia.

1978   5 December: Vietnam invades Cambodia en masse.

1982   13 November: The Vietnam Veterans Memorial, known as the Wall, is dedicated in Washington, D.C.

1986   General Eugene Tighe investigates the POW/MIA issue and concludes that there is a strong possibility that U.S. POWs are still being held in Southeast Asia. Under Vietnamese Communist Party General Secretary Nguyen Van Linh, a liberal economic policy is initiated in the consolidated Socialist Republic of Vietnam.

1989   The Socialist Republic of Vietnam withdraws its troops from Cambodia.

1991   October 1991–1992: The U.S. Senate Select Committee investigates the POW/MIA issue, focusing on U.S. government policy and practices, the question of possible American POWs being held after the war, and the declassification of relevant documents.

July: A *Wall Street Journal* poll indicates that 69 percent of Americans believe that POWs were left in Southeast Asia and may still be alive.

December: The dissolution of the Soviet Union impels Vietnam to renew links with the United States because the Soviet Union no longer counterbalances China in world-power politics.

5 December: The McCain Amendment to the National Defense Authorization Act requires the U.S. Department of Defense to declassify information relating to the location, treatment, or condition of any Vietnam-era POW/MIA.

1992   2 April: U.S. President George Bush lifts Vietnam trade embargo restrictions on telecommunications and on nongovernmental humanitarian and nonprofit organizations; he also allows the sale of products for basic human needs. The Senate Select Committee issues the final report of its investigation. It does not rule out the possibility that live Americans are being held in Southeast Asia, and it finds that government agencies responsible for dealing with the POW/MIA issue had

1992
cont.
been variously effective or ineffective depending on the priorities set and resources allocated by the presidential administration in charge at the time.

The Vietnamese alter the constitution of the Socialist Republic of Vietnam to allow foreign investment in their country.

22 July: President Bush signs Executive Order 12812, providing for expedited declassification of documents relating to POW/MIAs.

14 December: Bush allows U.S. companies to open offices, sign contracts, and do feasibility studies in Vietnam.

1993
The Department of Defense creates the Prisoner of War/Missing in Action Office (DPMO) to join four disparate offices working on the POW/MIA issue.

May: In Cambodia, elections sponsored by the United Nations result in a national coalition government.

10 June: U.S. President Bill Clinton issues Presidential Decision Directive NSC-8, which stipulates that all POW/MIA pertinent documents and files should be declassified and sent to the Library of Congress by the end of 1993.

24 September: Prince Sihanouk is restored as king of Cambodia.

1994
20 January: A CBS-*New York Times* poll indicates that "more than half of Americans believe MIAs are alive in Vietnam."

4 February: With U.S. Senate approval, President Clinton lifts the remaining trade embargo against Vietnam.

1995
27 January: The United States and Vietnam agree to open liaison offices in each country's capital.

June: Congressional hearings on the Vietnam POW/MIA issue take place to assess the effectiveness of

DPMO operations and the level of cooperation by the Vietnamese, Laotians, and Cambodians.

11 July: President Clinton announces the establishment of full diplomatic relations between the United States and Vietnam.

5 August: U.S. Secretary of State Warren Christopher opens the American embassy in Hanoi.

November: Congressional hearings on the POW/MIA issue take place to probe the possibility that POWs from the Korean and Vietnam wars were sent to the Soviet Union.

1996   Through the fiscal year 1996 National Defense Authorization Act, Congress overhauls the procedures that the Department of Defense (DoD) uses to determine the status of missing service members, DoD civilian employees, and DoD contractor personnel.

March: U.S. soldier Mateo Sabog, who since 1979 had been presumed killed in Vietnam, reappears live and well, having been living quietly for years in Georgia under an assumed name.

10 April: The U.S. Senate confirms Douglas "Pete" Peterson, a former Vietnam POW, as the first U.S. ambassador to Vietnam since the ending of the Vietnam conflict and the first ever to be posted to Hanoi. Le Bang is confirmed to be the Vietnamese ambassador to the United States.

1998   1 April: The official number of U.S. service members unaccounted for from the conflict in Southeast Asia is 2,093.

9 April: The National Prisoner of War Museum at the Andersonville National Historic Site in Georgia is dedicated. Former American POWs alive number approximately 56,000.

Biographical Sketches 3

Richard T. Childress

A specialist in East Asian Affairs, Richard Childress has worked at many levels on issues involving Vietnam. As a member of the U.S. military, in 1968 Childress served as a district senior advisor to the South Vietnamese in the Mekong Delta, but after a year he was posted to Germany for the remainder of the war. In the late 1970s, he served as a general staff officer in the Pentagon, where he prepared staff papers suggesting strategies to aid in a political settlement in Cambodia. In 1981, Childress became the director of Asian affairs at the White House National Security Council. During his eight years in this position, he conducted high-level negotiations with the Vietnamese on such subjects as the repatriation of stored remains of prisoners of war, the orderly departure program for Amerasians and Vietnamese wishing to emigrate, and the withdrawal of Vietnamese forces from Cambodia.

After Childress left government service in 1989, he became a consultant on Asian affairs and foreign investment. He continued his efforts to aid the Vietnamese by helping with the National League of Families and humanitarian organizations that bring relief to suffering Vietnamese and help to resettle refugees. He was vocal in his opposition to the lifting of the trade embargo and the normalization of relations with the Socialist Republic of Vietnam.

Ann Mills Griffiths

Since August 1978, Ann Mills Griffiths has been the executive director of the National League of Families of American Prisoners and Missing in Southeast Asia, a nonprofit, charitable organization made up of family members of POWs and MIAs dedicated to securing the return of all prisoners, the fullest possible account for the missing, and the return of remains of those who died while serving the United States in Southeast Asia. Griffiths's brother, Lt. Commander James B. Mills, has been missing since 21 September 1966, when the Navy F4C on which he served was lost on a night mission over North Vietnam.

Prior to assuming her position as executive director, Griffiths was an elected member of the league's board of directors for four years, serving as legislative chairwoman. She also played an active role in the U.S. government's POW/MIA Interagency Group (1980–1992), where she represented the families' views on development of official policy on this issue.

An expert on the POW/MIA issue, Griffiths frequently meets with senior administration officials and members of Congress, testifies before congressional committees, addresses national and international audiences, participates in appropriate policy seminars, publishes articles and newsletters, and appears on network and cable television programs. She plans the league's yearly convention, and with her staff of state and regional volunteers, she coordinates a nationwide public awareness campaign.

John Kerry (b. 1943)

U.S. Senator John Kerry of Massachusetts has been an advisor to President Clinton on foreign affairs and pushed for normalizing relations with Vietnam. Prior to entering the political arena, Kerry served (1968–1969) in the U.S. Navy as a small-craft commander in the Mekong Delta during the Vietnam conflict. For his service he received many decorations, among them the Silver Star, the Bronze Star, three awards of the Purple Heart, and the Presidential Unit Citation for Extraordinary Heroism. After returning from Vietnam, Kerry protested the war as the leader of the Vietnam

Veterans Against the War and earned a law degree. He was elected to the Senate in 1986.

With Senator Robert C. Smith, Kerry cochaired the Senate Select Committee on POW/MIA Affairs (1991–1993), which concluded that "there is no compelling evidence that any American remains alive in captivity in Southeast Asia." Though Kerry admitted that the issue has not been put to rest, he and Senator John McCain sponsored an amendment in 1994 to lift the longtime trade embargo with Vietnam. Kerry supported the normalization of relations with Vietnam as a way to gain access and move toward full resolution of the MIA issue. Kerry has also served as a commissioner of the Joint U.S.-Russia Commission on POW/MIA Affairs, which includes work on Korean and World War II MIA cases.

John McCain (b. 1936)

John McCain was the most seriously injured American prisoner to survive the Vietnam war and has since been active in MIA issues as a U.S. congressman. McCain came from a long line of military leaders and dreamed of commanding an aircraft carrier. This goal was thwarted, however, when during the Vietnam war he flew low-altitude bombing runs over North Vietnam and was shot down by a surface-to-air missile in 1967. After ejecting from his aircraft, he fell into a lake, breaking both arms and one leg. He was beaten by Vietnamese onlookers before being taken prisoner. After some time in a hospital, McCain spent three years in solitary confinement at a prison camp. Later he was housed with other American POWs. Despite his being the son of an admiral, he suffered the same deprivations and torture as other prisoners. McCain returned to the United States during Operation Homecoming. For his service in Vietnam, he was much decorated, earning the Silver Star, the Legion of Merit, and the Distinguished Flying Cross, among other medals.

After recuperating, McCain resumed his duties as a navy pilot, but by 1981 he realized that his military career was over because he could no longer pass the rigorous flight physical. He served as a Navy Senate liaison officer for several years and then retired from the military, settled in Arizona, and ran for public office. While a representative and then senator in the U.S. Congress, he has actively pressed the U.S. and Vietnamese governments to make progress on resolving the MIA issue.

As a senator, McCain sat on the Senate Armed Services Committee and in 1991 sponsored an amendment to the Defense Authorization Act of 1991 (authorizing funds for fiscal year 1992–1993) that was passed and later became known as the McCain Act. This act required the Department of Defense to declassify information pertaining to POWs and MIAs and make it available to the public through the Library of Congress. In 1995, McCain supported President Clinton's decision to lift the trade embargo and resume diplomatic relations with Vietnam.

Ross Perot (b. 1930)

Self-made billionaire businessman Ross Perot has put his resources behind efforts to aid and honor American soldiers who fought in Southeast Asia. After serving in the U.S. Navy, from which he was discharged in 1957, Perot went on to work for International Business Machines and founded Electronic Data Systems, which he staffed largely with Vietnam veterans. After supporting Richard Nixon in the 1968 presidential campaign, Perot made proposals to boost Nixon's popularity and founded the United We Stand pressure group to counteract anti-Vietnam sentiment among the public.

In 1969, Perot orchestrated the airlifting of Christmas dinners for the 1,420 American prisoners of war then known to be held by the North Vietnamese. He chartered two Braniff jets, one of which was then loaded with packaged dinners, clothing, medicines, and presents. The other held the wives and children of prisoners, who flew to Paris in efforts to speak with Vietnamese representatives. Although the North Vietnamese would not let the plane laden with supplies land, the publicity generated worldwide by the attempt is thought to have greatly improved the conditions under which the prisoners were held. After Operation Homecoming, Perot underwrote a ticker-tape parade in San Francisco for the returned prisoners.

In 1970, Perot became convinced that prisoners who were not on official lists were being held in caves in Laos, and in 1973 he paid for a small team of men to go to Laos and search for live Americans. None were found. Perot may have bankrolled other such missions, and in 1986 he financially supported an effort to obtain a videotape that allegedly showed live prisoners. The effort was unsuccessful. That same year Perot was given access to classified DIA files on the POW/MIA issue and spent several months studying them to see if the issue had been ignored for political reasons.

Convinced that conspiracies and cover-ups were obscuring the issue, Perot went to Vietnam, unauthorized, in 1987 and spoke with Vietnamese officials. Although Perot denies it, some people believe that in the following years he offered to ransom the prisoners for as much as $1 million each.

Perot was the single largest financial contributor to the Vietnam Veterans Memorial in Washington, D.C., and had some influence on its eventual design.

Pete Peterson (b. 1935)

In 1996, Douglas "Pete" Peterson became the first American ambassador to Vietnam since the Vietnam conflict and the first ambassador ever to be posted to Hanoi. The son of a Postal Service worker, Peterson grew up in the Midwest. In 1954, he dropped out of college to join the Air Force and later served in Vietnam. During a bombing run over North Vietnam in 1966, Peterson's aircraft was shot down by a surface-to-air missile, and he was taken prisoner by the North Vietnamese. He spent six and a half years as a prisoner of war in Hanoi. After his release during Operation Homecoming in 1973, Peterson remained in the Air Force until his retirement in 1981.

In the private sector, Peterson started several businesses, including a construction company in Tampa, Florida, and a computer company in Marianna, Florida. He also worked as a counselor for troubled teenagers. But by 1990, Peterson was looking for a new challenge. Running as a moderate Democrat, he was elected to the U.S. House of Representatives. During his three consecutive terms in the House, Peterson was active in supporting veterans programs. After his wife died in 1995, he decided not to run for reelection. Yet he soon found himself being tapped for the ambassadorship to Vietnam.

Mateo Sabog (b. 1921)

U.S. Army Master Sergeant Mateo Sabog, who had been listed as killed in action and whose name is on the Vietnam Veterans Memorial in Washington, D.C., was discovered in 1996 to have been living in Georgia for ten years under the assumed name of Bobby Fernandez and working jobs that paid in cash. Sergeant

Sabog was thought to have left Vietnam for Fort Bragg on his normal rotation date in February 1970, but there was no proof that he was ever on the plane. Because the authorities at Fort Bragg were not expecting him due to an administrative glitch, the army didn't know he was missing until 1973, when his family began making inquiries. The army initially listed Sergeant Sabog as a deserter; however, Sabog's family contested that classification. After an army investigation and an FBI search proved to be unfruitful, his status was changed in 1979 to missing with a presumptive finding of death as of 1970. At the time of his disappearance, Sabog had served 24 years on active duty, with a spotless record that included a 1-year tour in Vietnam.

In 1995, the POW/MIA Office notified Sabog's brother that Sabog's remains had been recovered and that it was about to conduct a DNA analysis to confirm his identity. Before the test could be done, however, Sabog showed up at a Social Security Administration office in Georgia to apply for benefits. When the computer records indicated the application was being made in the name of a man who was officially classified as dead, fingerprints were compared, and they proved Sabog was who he claimed to be. In March 1996, Sabog was returned to active duty so he could be admitted to the Eisenhower Army Medical Center, Fort Gordon, Georgia, for evaluation and any needed medical treatment. Afterward, he was honorably discharged from the U.S. Army. He returned to Hawaii to live, saying little to solve the mystery of his whereabouts for the decade between his service in Vietnam and his appearance in Georgia. This mystery has spawned speculation that Sabog was held prisoner in Vietnam after the end of the war.

Patricia B. Skelly (b. 1946)

Patty Skelly is the chairwoman of the board of directors of Task Force Omega, Inc., a POW and MIA organization whose goal is the return of all Americans from Southeast Asia, both alive and dead.

Skelly first became aware of and interested in the POW/MIA issue in 1966 as a military wife while her husband was stationed in Vietnam. Her interest increased in 1972 when a close high school friend became missing in action.

Skelly has been actively involved in the POW/MIA issue since 1975 when, as a concerned citizen, she was appointed Minnesota state coordinator of the National League of Families of

American Prisoners and Missing in Southeast Asia. She served in this capacity until 1980. From 1980 to 1983, she served as Minnesota state coordinator for the National Forget-Me-Nots Association for POW/MIAs. During 1981 and 1982, Skelly was also the POW/MIA state chairperson for the Department of Minnesota Veterans of Foreign Wars Auxiliary.

In 1978, Skelly became involved in research and analysis of POW/MIA material, especially the live POW reports from refugees and other sources, due to the many discrepancies that were becoming apparent in U.S. government documents. The results of this work have been, and will continue to be, shared with other individuals and organizations involved in the POW issue. Skelly is considered to be an expert on POW/MIA matters.

In 1983, the need for an organization in the private sector devoted to research and analysis of the POW/MIA issue with a membership open to all Americans became apparent. As one of the founders of Task Force Omega, Inc., Skelly has continued to seek information on the fate of MIAs.

During her many years of involvement in the POW issue, Skelly has met with numerous government officials, including congressmen, senators, State Department officials, and Defense Department officials, and testified before Congress. She works with members of many other veterans and POW/MIA organizations.

Robert C. Smith (b. 1941)

U.S. senator from New Jersey Robert C. Smith has made POW/MIA advocacy one of his primary political causes. As a congressman in the U.S. House of Representatives from 1985 to 1990, Smith, a former U.S. Navy enlisted man who was on active duty in Vietnam (1966–1967), served on the Armed Services and Veterans Affairs Committees as well as the House Task Force on POW/MIAs.

While in the Senate, Smith served as chairman of the Vietnam War Working Group of the Joint U.S./Russia Commission on POW/MIAs. He also traveled to Korea to talk with officials about Korean War POWs and repatriated the remains of several servicemen. Smith introduced legislation that ultimately formed the Senate Select Committee on POW/MIA Affairs, which he cochaired for 18 months (1991–1993) with Senator John Kerry. Although Smith signed off on the Senate Select Committee's report that "no compelling evidence" of POWs still alive in Southeast Asia exists, he subsequently took up the cause of families of

MIAs, insisting that every possibility must be thoroughly investigated.

To that end, Smith has played the watchdog over the activities of the DPMO, the military, and the Bush and Clinton administrations. During congressional hearings and from the Senate floor, he vigorously opposed all efforts to normalize relations with Vietnam, maintaining that the Vietnamese government continues to withhold the remains of U.S. servicemen and information about their fate.

John Vessey (b. 1922)

In the U.S. Army, John Vessey served his country for 46 years, rising from the rank of private to eventually become the chairman of the Joint Chiefs of Staff. After retiring from the army, he acted as a special emissary to Hanoi under three presidents (Reagan, Bush, and Clinton). His mission was to negotiate with the government of Vietnam about accounting for prisoners of war possibly still held by the Vietnamese and about the repatriation of the remains of those servicemen killed in action whose bodies had not been recovered. In 1987, he made the first of his six missions to Hanoi.

Vessey testified during the Senate Select Committee's investigation in 1992 and at other congressional hearings pertaining to the MIA issue. He played an important role in establishing the Joint Task Force–Full Accounting (JTF-FA) and the full-time POW/MIA Office in Hanoi. In 1993, Vessey led a team that investigated POW information that had been received from archives of the former Soviet Union and Vietnam's political directorate.

James W. Wold (b. 1932)

Former Deputy Assistant Secretary of Defense for Prisoners of War/Missing in Action James W. Wold is a veteran of the Vietnam war. While in Southeast Asia, he flew 241 combat missions, including search-and-rescue missions over Vietnam and Laos for the U.S. Army Special Forces Teams and other ground tactical units. A highly decorated veteran, he was awarded the Legion of Merit, the Distinguished Flying Cross, the Air Medal, and the Bronze Star, among other awards.

Rising to the rank of brigadier general at age 41, Wold became one of the youngest generals at that time. He commanded

Detachment A of the 56th Special Operations Wing in Vietnam. In the Pentagon, he was chief of the Colonel's Division at Headquarters, U.S. Air Force, and as such was personally involved in the repatriation of U.S. prisoners of war released during Operation Homecoming in 1973. After 26 years of duty with the Air Force, Wold retired and as a civilian ran a successful general law practice in Cooperstown, North Dakota, before being chosen to head the DPMO.

Statistics and Documents 4

L iterature on the MIA topic, which is abundant, varies greatly in quality. Among the works on the subject are the millions of words recorded under oath and published as testimony before congressional panels; the arguments of MIA activists published in books, newsletters, and Internet sites; and declassified government intelligence reports and other primary-source material. Perhaps the most interesting words would be found in the unpublished sources—such as testimony made in executive session at congressional hearings, or the many documents that remain classified for security reasons despite presidential directives to declassify relevant documents.

This chapter presents some of the published wealth of information about the MIA issue as it relates to the Vietnam conflict and, more broadly, to warfare in the modern era: general facts and statistics relating to MIAs, excerpts from pertinent documents from a variety of congressional sources, excerpts from the 1949 Convention for the Amelioration of the Condition of the Wounded and Sick in Armed Forces in the Field and the 1973 Paris Peace Accords, the secret Nixon letter, Section 569 of the 1996 National Defense Authorization Act, and speeches made by public officials.

Statistics

Readers may draw several conclusions from the data presented in this chapter. First, compared with the statistics from previous wars, the numbers of prisoners of war and missing service members from the Vietnam conflict are very small (see Tables 4.1 and 4.2). This is so in part because of improved U.S. search-and-rescue operations. Despite the small numbers, these MIAs, whose families represent every state in the union (see Table 4.3), have dominated U.S. foreign policy toward Vietnam for over 20 years. The dominance of the MIAs represents the larger issue of how the U.S. government treats and will treat its military forces in times of conflict. More than 3 million American veterans have a stake in this issue, as do countless future service members.

Table 4.1.
American Casualties and Prisoners

	Killed	Captured and Held	Died in Captivity	Released from Captivity	MIAs
World War I	116,516	4,120	147	3,973	N/A
World War II	405,349	130,201	14,072	116,129	78,750
Korean War	54,246	7,140	2,701	4,418	8,300
Vietnam War	58,022	766	106	653	2,583

Sources: U.S. Department of Defense 1978, Table 28.2, 1981; Department of Defense POW/MIA Office, cited in National League of Families of American Prisoners and Missing in Southeast Asia 1997; Olson 1993, 476; Veterans Administration 1980, 10.

Table 4.2.
Americans Unaccounted for in Southeast Asia
(As of 24 October 1995)

Country of loss	POW/MIA	KIA/BNR	Total
North Vietnam	336	257	593
South Vietnam	430	591	1,021
Laos	293	178	471
Cambodia	36	41	77
China	6	2	8
Total	1,101	1,069	2,170

Note: Total includes 41 civilians: 34 POW/MIA, 7 KIA/BNR.
Source: U.S. House of Representatives, *United States and Vietnamese Government Knowledge and Accountability for U.S. POW/MIAs* 1996c, 85.

Table 4.3.
Listing by State of Unaccounted-for Servicemen
(As of 1 July 1990)

Alabama: 42	New Hampshire: 10
Alaska: 2	New Jersey: 61
Arizona: 23	New Mexico: 17
Arkansas: 26	New York: 144
California: 230	North Carolina: 57
Colorado: 41	North Dakota: 16
Connecticut: 37	Ohio: 117
Delaware: 5	Oklahoma: 47
District of Columbia: 9	Oregon: 44
Florida: 77	Pennsylvania: 115
Georgia: 44	Rhode Island: 9
Hawaii: 10	South Carolina: 30
Idaho: 10	South Dakota: 9
Illinois: 95	Tennessee: 42
Indiana: 66	Texas: 146
Iowa: 38	Utah: 19
Kansas: 35	Vermont: 4
Kentucky: 21	Virginia: 55
Louisiana: 30	Washington: 57
Maine: 17	West Virginia: 24
Maryland: 36	Wisconsin: 37
Massachusetts: 57	Wyoming: 6
Michigan: 73	Puerto Rico: 2
Minnesota: 41	Virgin Islands: 1
Mississippi: 18	Other: 7
Missouri: 49	
Montana: 21	
Nebraska: 22	
Nevada: 9	

Source: U.S. Department of Defense 1990, 4.

Second, the uncertain nature of the numbers of POWs and MIAs has caused controversy dating to the negotiations that led to the signing of the Paris Peace Accords. Because the Nixon administration used the POW and MIA issue to garner public support for a continued war effort, it is reasonable to believe that the numbers cited by the administration were higher than the actual number of live prisoners or MIAs believed to exist. Despite its best intelligence efforts, the U.S. military could not determine the exact number of prisoners held by the North Vietnamese or Pathet Lao forces. Yet as of December 1972, the Pentagon listed 620 Americans as prisoners of war (Office of Assistant Secretary of Defense 1994, 3). Moreover, because the North Vietnamese did not adhere to the 1949 Geneva Convention, which mandated that captors keep and share records of their prisoners with their enemy, U.S. leaders had only military intelligence data on which to base estimates of POW numbers under Communist control. Indeed, the North Vietnamese insisted that both sides produce complete lists of military and civilian prisoners only on the day the Paris Peace Accords would be signed.

Statistics concerning MIAs are subject to differences in opinion and are very confusing. They have been in a constant state of flux as cases are reviewed and status changed. During the war, individual military units and various other groups (including the Defense Intelligence Agency, Pentagon, and State Department) kept their own casualty records and POW and MIA records. After 1973, the Pentagon compiled a list that grouped together both those service members thought to have been KIA/BNR and those who were MIA—about 2,000 in all. Approximately half of these men were previously thought to actually be KIA/BNR. Thus, priority has been given to resolving what have been called discrepancy cases, that is, those cases in which U.S. intelligence data support the contention that at one time an American prisoner was known to be in North Vietnamese or Pathet Lao hands but was not returned to U.S. control at war's end.

In some instances, family members of MIAs have contested status review changes and what they considered to be inconclusive evidence used to identify remains as being those of a particular serviceman (e.g., Peter Cressman; see testimony of Patrick Cressman below).Therefore, these families cast doubt on the accuracy of government statistics. In addition, the number of remains repatriated and positively identified has been used by various administrations and observers of the issue to measure the success of U.S. government efforts to resolve the POW/MIA issue. Before 1988, most remains were unilaterally repatriated by the Vietnamese, and until 1991, when Vietnam discontinued the repatriation of stored remains, almost 65 percent showed signs of having been stored (Dole 1995, S218).

It used to be that until the Central Identification Laboratory in Hawaii (CILHI) positively identified remains as being those of a missing serviceman, no public announcement was made. However, after President Clinton took office, announcements of remains repatriations were made prior to positive identification, a practice that inflated the actual number of cases resolved at the time of the announcement. Despite any public relations maneuvers, a trend is clear (see Table 4.4). After the normalization negotiations by the Carter administration broke down when the Socialist Republic of Vietnam invaded Cambodia in 1978, remains repatriations stopped until President Reagan revisited the issue with renewed vigor. During phases of the Bush administration "road map" and the efforts to normalize relations with Vietnam by President Clinton, a steady but small flow of remains has been repatriated unilaterally and through joint U.S.-Vietnamese excavation efforts. On its Web site the DPMO announces the resolution of MIA cases as they are concluded.

Table 4.4.
Americans Accounted for since 1974

1974–1975	Postwar years	28
1976–1978	US/SRV normalization negotiations	47
1979–1980	US/SRV talks break down	1
1981–1984	First Reagan administration	23
1985–1988	Second Reagan administration	153
1989–1992	Bush administration	108
1993–1996	First Clinton administration	121
1997	Second Clinton administration	35

Note: Unilateral SRV repatriations of remains have accounted for the vast majority of the 516. The breakdown by country of the 516 is Vietnam, 387; Laos, 120; China, 2; and Cambodia, 7.
Sources: Department of Defense POW/MIA Office, cited in National League of Families of American Prisoners and Missing in Southeast Asia 1997; U.S. Department of Defense POW/MIA Newsletter 1 April 1998, n.p.

Third, by the time of the 1973 cease-fire there had been many sightings of live POWs in Laos, and since then the DIA and its successor, the Department of Defense Prisoner of War/Missing in Action Office, have expended a great amount of time and effort investigating them. From 1975 until 31 October 1997, these offices received 1,867 firsthand live-sighting reports in Indochina; 1,781 (95 percent) have been resolved. Of the reports resolved, 1,249 (66 percent) were equated to Americans now accounted for (i.e., returned POWs, missionaries, or civilians detained for violating Vietnamese codes), 45 (3 percent) correlated to wartime sightings of military personnel or pre-1975 sightings of civilians still unaccounted for, and 487 (26 percent) were determined to be fabrications. The 86 (5 percent) unresolved firsthand reports are the focus of current analytical and collection efforts: 77 of these (4 percent) are reports of Americans sighted in a prisoners situation, and 9 (1 percent) are non-POW sightings.

Finally, much work remains to be done. After becoming head of the DPMO in 1993, Deputy Secretary of Defense (POW/MIA

Affairs) James W. Wold conducted a comprehensive review of outstanding POW/MIA cases. The review was a combined effort of JTF-FA, CILHI, and the DPMO. Their final report summarized file by file what had been done to date on a case and what the Vietnamese government had provided on the case; predicted whether the Vietnamese, Cambodian, or Laotian governments were likely to possess more information on a case; and decided whether to continue to actively investigate a case or defer it pending the receipt of further information. As of 30 November 1995, 1,476 cases (67 percent) of the total of unresolved MIA cases were marked for further pursuit. Of this total, 942 were in Vietnam, 470 in Laos, 61 in Cambodia, and 3 in China. Of the remaining 567 (33 percent) "we judge no actions by any government will result in the recovery of remains" due to the nature of the loss, such as explosions, aviators lost at sea, or remains buried in areas where the topography has changed significantly (U.S. Department of Defense Review 1996, 47).

In his study *Code-Name Bright Light*, George J. Veith (1998) describes the great efforts made by the U.S. military to rescue American prisoners of war. After the unsuccessful attempt by U.S. forces to rescue prisoners of war from the Son Tay prison camp in North Vietnam, the DRV (North Vietnamese Forces) are believed to have consolidated American prisoners in several prisons in Hanoi proper. At the time of the cease-fire, approximately 160,000 North Vietnamese Army troops were holding territory in South Vietnam, a situation that South Vietnamese President Nguyen Van Thieu viewed as a prelude to disaster—the eventual total takeover of the South after the withdrawal of U.S. troops. As history would prove, Thieu was correct.

Documents

Following are excerpts from the congressional testimony of three families of MIAs regarding individual cases that typify the types of experiences families have endured and the complaints they have voiced as a result. These are preceded by the military Code of Conduct, a version of the tap code used by American prisoners to communicate with each other across cells, and a riveting account of POW Eugene "Red" McDaniel's treatment at the hands of the North Vietnamese. Ted Guy's personal account of captivity and the reasoning behind his belief that American POWs could survive until the present finishes the section.

Code of Conduct for Members of the Armed Forces of the United States

In the wake of the Korean War, when fears of brainwashing of prisoners of war were high, a number of countries instituted code of conduct regulations to guide the behavior of prisoners of war. By an executive order on 17 August 1955, President Dwight D. Eisenhower created a code of conduct that emphasized divulging only "name, rank, and serial number." Eisenhower further directed that U.S. military personnel undertake training in interrogation resistance methods. When the experiences of prisoners undergoing torture at the hands of the North Vietnamese during the Vietnam war became known, President Jimmy Carter revised paragraph five. In 1977 he mandated that the word "required" be substituted for "bound" to allow prisoners suffering torture to tell more than their name, rank, service number, and date of birth without shame.

1. I am an American fighting man. I serve in forces which guard my country and our way of life. I am prepared to give my life in their defense.
2. I will never surrender of my own free will. If in command I will never surrender my men while they still have the means to resist.
3. If I am captured I will continue to resist by all means available. I will make every effort to escape and aid others to escape. I will accept neither parole nor special favors from the enemy.
4. If I become a prisoner of war, I will keep faith with my fellow prisoners. I will give no information or take part in any action which might be harmful to my comrades. If I am senior, I will take command. If not, I will obey the lawful orders of those appointed over me and will back them up in every way.
5. When questioned, should I become a prisoner, I am bound to give only name, rank, service number, and date of birth. I will evade answering further questions to the utmost of my ability. I will make no oral or written statements disloyal to my country and its allies or harmful to their cause.
6. I will never forget that I am an American fighting man, responsible for my actions, and dedicated to the principles which made my country free. I will trust in my God and in the United States of America.

Source: Hackworth and Sherman 1989, 360.

Tap Code

While in captivity, American prisoners devised communication systems, such as the tap code below, to maintain a chain of command, account for prisoners held, and boost morale. Although their Vietnamese captors decoded the system, they were unable to fully understand prisoner communications because the prisoners used slang and shorthand. The letter K was omitted to make the code mathematically symmetrical and because its function can be performed by the letter C or—in the case of plurals—by the letter X.

A	..	J	. .....	S	....
B	. ..	L	.... .	T	....
C	. ...	M	.... ..	U	....
D	. .....	N	....	V	....
E	. ...	O	....	W	....
F	... .	P	....	X	....
G	... ..	Q	..... .	Y	....
H	.. ...	R	....	Z	....
I	.. ....				

Source: Rowan 1973.

Personal Accounts of POWs and Their Families

"Escape and My Darkest Hour"

Navy pilot Captain Eugene "Red" McDaniel was shot down during his eighty-first combat mission in 1967. During his six years as a prisoner of the Vietnamese Communists in Hanoi, he endured brutal treatment, which he miraculously lived to describe in his book Scars and Stripes, *excerpted below.*

There were nights in the Annex when Bill Austin and I talked about escape. Sometimes it was not serious talk, but something merely to occupy our minds. We knew about the code of conduct according to which it was the prisoner's responsibility to escape when possible, but even beyond this the urge to make it out to freedom was often overpowering.

There had been an escape attempt in October 1967

when Air Force Captain George McKnight and Navy Lieutenant (jg) George Coker managed to get out. They floated down the Red River for about fifteen miles until they got chilled and decided to come out and try to go it on land. They were picked up immediately by Vietnamese civilians. The VC did not punish them, because both men said the attempt was out of desperation—they felt they would die in prison anyway so they might as well try it. After that, however, the camps were on very tight security. Later, in December 1969, there was yet another attempt— this one by Colonel Ben Purcell—but he did not get very far before being picked up. He was put in a punishment camp called Skid Row, because it was twice as bad as anything in the "normal" prison environment, reserved for those who were incorrigible.

As Austin and I talked about it in the Annex in late 1968, we weren't sure it was possible to make a successful breakout and make our way the hundreds of miles to the sea or to friendlies in the south. For one thing, Hanoi sat on the Red River Delta, where several hundred thousand people lived. Even if we did make it out of the compound, the people in Hanoi were dedicated to tracking down any foreigners and were promised $1,500 by the government for every American flyer they caught. It would be tough trying to get through that kind of "police force." What's more, the average North Vietnamese is about five feet two inches tall, with black hair, yellow-skinned, slant-eyed. How would I, at six feet three inches, now about 160 pounds, with red hair, fair-skinned, and round-eyed, fit into that population if I tried to use its traffic patterns to get to the sea?

And there was another problem, that of environment— the miles and miles of jungle, the thick, cruel bush that a man would have to negotiate to get out. To illustrate the nature of the land we were living in, Bill Austin killed a poisonous snake in our room one night, and in the morning when we emptied our toilet bowls, we hung it up on the fence. We watched that snake periodically through the door peephole, and in three hours it had been completely devoured by insects. Nothing was left of it—nothing. We knew the same thing could happen to us if we tried to make it 110 miles to the sea.

For all that, we still spent time thinking up ways to

break out of the prison itself. Austin and I managed, after weeks of labor, to get the inner metal doors off their hinges. It really would have been no problem to walk straight out the door if we wanted to. And then one night in the room next to us, where Captain Konrad Trautman and three others lived, the guard had forgotten to lock the door altogether. That was a rare occasion, but it happened, and the men there thought all night that unlocked door. In the morning Trautman pounded out [using the tap code] the message to us that the door had been open all night and then added, "But where to next?" That about summed it up. If we did get out, where then? There were too many obstacles to success. And, unless we had help from the outside, someone coming in with the means to jump us out of there completely, it would be suicide to try it.

Well, in May 1968, I was moved again--this time to the Zoo, which was directly behind the Annex. We had not known it was there until earlier that year, when Major Al Runyan came into the Annex from the Zoo. I moved over to a building in the Zoo called the Garage with Runyan and Ken Kleenor. The layout of this camp was the same as the Annex, except there were eight buildings instead of five. The Garage was directly behind the fifteen-foot wall separating the Zoo from the Annex, and, since we knew the layout of the Annex, we began to set up a communications system that would link the Zoo and the Annex, which up to now had not been established. I reasoned that, if we could bring the 118 men in the Zoo into contact with the 72 men in the Annex, we would have much more going for us.

The only way to get through the fifteen-foot wall that separated the two compounds, then, was to toss notes over the top or try to communicate through the ceiling vents of our rooms. We ran our "pony express" in the same way, but could not cross over to the Annex with that wall between us. Still we managed to make our communication work very well.

Meanwhile, the Paris Peace Talks had resulted in our getting our first food packages from home; they arrived in February 1969. A few good things were happening about then in the prison. More of our wounded people were being taken to the hospital. Too, we were getting a little more food, and better food too—sometimes bread with our soup now, sometimes some canned meat, sometimes candy.

The VC [Vietcong], however, were demanding that we make tapes telling of the "good" treatment. This most of us never did. During the early spring of 1969, unknown to me, a nine-man committee which included Major John Dramesi and Major Ed Atterbury was storing extra food rations the Vietnamese were giving us—the result again of pressure in the Paris Peace Talks—hiding them up in the attics of their rooms. There were, of course, sick people who would not eat, and these men used their rations to put into an emergency kit for a breakout.

I was to remember the night of May 10, 1969, when John Dramesi and Ed Atterbury broke out. They climbed out of Room 6 in the Annex, up through the ceiling vent, down the wall, across the compound, and over the wall near the Quiz Room. I didn't know it until the message was pounded out to the entire complex, and I sensed then that we were going to be in for it.

Sure enough. For three weeks the Vietnamese, in a kind of rage, took out twenty men to the torture room. Others were given the psychological approach and offered a deal: If they would talk and tell who was responsible for the escape, they would receive benefits in return. A lot of people lied under torture, telling the Vietnamese anything to get relief, and each man was forced to put his finger on someone else. They couldn't be blamed, for the news went around that Atterbury had been caught and had died in the torture, so everybody knew this was a serious matter, and a man had to look out for himself. Dramesi, too, was recaptured. I knew it was only a question of time before they put the finger on me, because I was right in the center of the communications system.

On the night of June 14, 1969, they came for us. I heard the key in the lock, and my stomach jumped as it always did at that sound. Somehow I knew this was going to be the big one—that what I had up to now was kid stuff. They took Al Runyan out first, because he was the senior man in the room; then they came for Major Ken Fleenor. Finally they came and got me. I was told to take my drinking cup and suit up. The guard's bayonet prodded me along past the two rooms where Al and Ken had been taken, until we reached the last room in the building called the Chicken Coop.

I looked around the room. It had not been swept, and it was very dusty. It was lighted by a fifteen-watt bulb, which dangled from the ceiling like a piece of rotten fruit left behind in a tree. The switch to it was simply two wires hooked together. The door to the room had a blue curtain hanging over it, and the bottom was draped to about six inches from the floor. The windows had glass—the only glass in Vietnam, I supposed—but that didn't help, because boards were nailed over the glass, so there was no light coming in.

Now I sat there in that room, which was like a coffin; they put leg irons on me like loop-shaped U-bolts, with the large leg iron in between serving as the eyes of the U-bolts intertwining both ankles, which was to make sure I did not escape. I started yelling "Bao cao!" [interpreter] right away; I wanted an interpreter so I could ask why I was being put into leg irons, since the escape was not from my building. But they ignored it. They continued putting on the irons and I kept yelling "Bao cao"—again to no avail. Then my arms were locked behind my back with wrist irons, very tight. They were turning it on this time, rougher than they'd ever been. Somehow I sensed that this was the point I had been heading for all along; this was the crucial test to my own faith, to the sense of optimism I had tried to maintain all these months. Now it was going to be a battle between them and me; one of us was going to come out of this on top, and there seemed little hope that I could, that I would ever live through it. Yet I knew I had to try, to use all I had within me to resist.

Finally an officer came in, and I recognized The Rat—not Sweetpea this time, as I had expected. I knew for sure, then, that the interrogation was going to be serious. I looked up at his sharply defined features, long hair, pointed nose, dark-skinned face, and that belt wrapped around his middle; he had on rubber tire sandals. Joining him was Hanoi Fats, who had the extra thirty pounds to use to throw us around if we needed it. There was one other guard, fairly new to me, whom we called Switches, a man with a hating look. He would put the torture on when it was time, of that I was sure.

The Rat began with, "What building do you live in?"

"I don't know the name," I said. "It doesn't have a name."

"Who is the senior ranking officer in the camp?"

"I don't know." He unloosed the irons that held my hands behind my back and made me drop my pants onto

the floor, and I got down spread-eagled under his command, with my bare buttocks exposed. Then Switches took
a rubber fan belt about three and a half feet long, an inch in
diameter, which had a knot in one end to get a better grip,
and he began to strike me across the bare buttocks very
hard. I winced, with the pain of it. Five times he did it, then
stopped. I was allowed to pull up my pants, and again The
Rat started the same questions. I told him again I didn't
know.

Now he forced me to kneel down with my arms over
my head, my wrists locked in those irons again. I was told
to stay that way, not to let my arms drop at all. Well, for the
first hour it wasn't bad; after that, however, my arms began
to tire, and I let them relax a bit. Immediately a guard
behind saw my arms waver, and he hit me hard across the
back of the head. I put my arms up straight again, but now
it was as if fifty-pound sacks were hanging on them.

This went on all day Saturday. At three or so I couldn't
hold those arms up any longer, and I let them drop. A
guard jumped in and hit me with a karate chop on the back
of the neck, then a few more times around the head. I fell
over. Then Hanoi Fats came in and tied me up again, this
time wrapping ropes around my hands, then hiking them
up to the back of my head, tying my arms between the
elbows and the shoulder sockets so that I could not drop
them. I had no choice but to hold them up then.

The rest of the day it was the same treatment. If my
arms relaxed even an inch, Hanoi Fats would let me have
it. By now I had a cut on my mouth that had begun to
bleed. Finally an officer came in, and I recognized
Soft-Soap Fairy or The Mystery Man, as we sometimes
called him. I yelled at him a little bit. He asked the guard
what was wrong, and I showed him the blood I was
spitting on the floor. But he took no action. I stayed in that
kneeling position with my arms over my head until nine
o'clock that night. At nine they let me sit on a stool in the
middle of the room. I did not sleep at all that night. The
stool had one short leg, one side of the seat was raised
higher than the other, only because it was a poor piece of
carpentry. One of my rubber sandals had fallen a few feet
in front of me. Behind the table in front of me sat The Rat,
watching me coldly. When I asked questions, he beat me
with that rubber sandal across my face.

So the night passed on that stool, and I tried to ignore the pulsing pain in my face, my mouth, my arms, my shoulders, my legs. The next morning around five, WTG ["World's Tallest Gook," one of the guards] came in and gave me a cigarette and a light. He gave me about ten minutes to smoke it. When I finished I threw the butt into a corner.

Then The Rat came in. "Time for more questions," he said shortly. "If you answer my questions correctly, then I can go down and spend Sunday on the lakes, which I would much rather do. If you do not answer, then I don't care. We'll stay here until you do answer."

So I went through the same painful routine all through Sunday, until nine o'clock that night. I was beaten with a fan belt on ten different occasions on Sunday during the day, about five licks each time. Then at nine it was back on the stool again, the ropes and those heavy irons on my legs beginning to drag on me.

Monday was the same—still no sleep. Came Tuesday and I saw that my knees were infected from kneeling so long on them, and the skin around my Achilles tendons was oozing pus too from where the leg irons rubbed. On Tuesday two guards came in; one I recognized as The Slug, and the other was WTG. It was their day to stay with me and make sure I kept my arms over my head while kneeling down. Each time I should drop my arms after hours of holding them up, they would beat me around the shoulders with a bamboo stick.

Then I began to have hallucinations.

"Who is on the Zoo escape committee?" The Rat kept asking me.

I tried to tell him there was no such committee in the Zoo. But others who had been tortured already had told them the committee existed. And so I went on all day—and they started it all over again. I would not give them the escape committee—I felt I had to protect Larry Guarino, Ted Kopfman, and J. J. Connell, who were already suspected by the Vietnamese. I simply would not put the finger on them. But I realized by now that the others who had been tortured had probably told them that we had organized the escape out of the Annex, and this was the confession that they wanted to hear.

Finally, unable to bear the agony of it any longer, I

decided to tell them that my room in the Zoo was the source of the escape committee. I had to tell them something, anything, to get some relief. By now I couldn't think, I was hallucinating. I had had no sleep, hardly any food. The pain in my shoulders and my legs was a fire shooting through my body. My face was beaten to a pulp, my mouth swollen. So I said my room in the Zoo was the source of the escape.

"Who else is going to escape?" The Rat barked at me.

"I am."

"What are you going to take with you?"

"An escape kit . . ."

"What is in it?"

"Rice, iodine, bread, canned meat," I mumbled through swollen lips.

"Where do you have that supply hidden?"

"In the back of the bath behind the Garage."

So they stopped the interrogation and went to the bath and tore the bath apart looking for the escape supplies. They found nothing. They came back, and I knew they knew I was lying. But I at least had the hour to get some relief.

"You lied," The Rat said flatly. I waited to get hit, but instead he asked, "How do you plan to go out?"

"Through the front door, how else?" I said, and I remembered how Bill Austin and I had taken off that inner door once.

"How do you plan to get through the door?"

"I'll knock it down. . . ."

They stopped again and went out right away and put a large metal hasp on the door to reinforce it. They did this to all the doors, but only on mine did they put a heavy four-by-four beam in addition, which was to be called by the other prisoners "McDaniel's Escape Bar."

But they weren't satisfied with the information I had given them. They kept at me, trying to drag more from me. On the fifth night I thought I could sleep. WTG had said, "Five night sleep," and that apparently had been the policy. But on the fifth night I still was not allowed to sleep. In fact, I was forced to kneel down until past nine o'clock that night, and the officer stayed longer than usual. By then I was running a high temperature of 104 to 105 degrees from the infections in my knees and the oozing wounds on the backs of my heels from those irons. So much came out of those wounds that whenever I moved around in that small room a trail of pus would be left behind along the floor.

On Wednesday the same treatment, beginning with the "setup" cigarette at 5:00 A.M. Then, right on schedule, ten minutes later The Rat came in, and the questions began, the same ones.

By now I had begun competing with the guards in the room, Slug and WTG. In all of the exhaustion, when my mind wouldn't function, when the pain beat a steady rhythm throughout my body, I still rose to the occasion with an insane desire to beat them. I was determined to stand with my arms higher and longer than they expected. It was foolish, because it was only wearing me down more, the very thing they wanted. So why did I fight them, why was the game spirit coming up again in such a ridiculous place and in such a setting?

[McDaniel describes a high school coach who drilled into him the need to win.] "To win isn't everything—actually, it's the only thing." Every game, every challenge of life was built around that statement for me. The military picked it up and socked it home. You flew to win against the opposing elements of empty space; you flew to win in combat. The American way was to win. "I never saw an American who lost and laughed," General George S. Patton said once. I wouldn't lose. I couldn't lose. I had been drilled to win.

But "winning" was not the issue anymore to many people. Many people had said that about the war; we'd heard it enough on the "box" over Radio Hanoi. It's not a "win" you're after; a "draw" might do. I knew that every-body was talking about the hopelessness of winning this war in Vietnam . . . maybe they were right.

But I couldn't turn if off that easily. I had to hang on to the winning, I had to compete, or I was dead in this place. Frank Mock's [the coach's] drive to make me want to win had to count now in the biggest battle of my life. And yet I knew, too, as Grantland Rice had put it, "He marks—not that you won or lost—but how you played the game." Frank Mock had inspired me to win but also to put character into the game, to put all I had into it regardless of the outcome. . . . But how do I play this "game"—locked in the ropes, bleeding, burning up with fever, my life oozing out of broken flesh, my mind so far gone I couldn't concentrate on their questions, could hardly hear them? Yet all of my discipline of the past told me I had to try.

Were there no winners? If that were true, if that philosophy were right, then what was I fighting for in this room? Why didn't I just roll over and die, let it happen?

Pride? Yes, I had it. I played the game and I was proud to be playing. All my life I had found pride in accomplishment, in winning, sometimes even just being out there and competing. I was proud to fly formation with my fellow pilots. I was proud to fly for the United States. I had caught the team spirit—loyalty to cause and country—early in life. Perhaps I wasn't so proud of letting go with bombs on targets; no combat man is. There is nothing to be proud of in inflicting pain and death. But there is something to be proud of in aiding the cause of a people who have no one to care, who have to stand alone. I was proud to be a part of my country's concern to do that. Regardless of what history would say, I could not deny my beliefs—I threw myself into the Vietnam war, because I believed, like my government, in the rightfulness of it and I was proud to play my part. Even if that cause is not a popular one, a man had to take his stand for something sometime. My Communist captors, the VC, had their cause and they were out to win—to tell them there is no winning in this war would bring laughter. Tell them to learn to be "graceful losers"; they wouldn't understand the statement. For all the bombing we did, they still had squadrons of rolling trucks full of soldiers and equipment. For all our superior fire power, they could still storm barbed wire in human waves. For them, winning was everything—and *their* pride was in even dying to prove it.

So what did I have left in this bloody hole I was in, being slowly reduced to a broken, beat-up Navy flier? If I were going to live through this, all that had been given to me through my life that counted at all had to be dredged up now. Only the armchair strategists in their wall-to-wall comfort could afford to philosophize about winning or not winning. Pain has a way of focusing for a man, giving him recall of what really matters. With death knocking at the door, a man has time for only one thing: to go out with some kind of honor.

Around noon The Slug was still watching me. I knew

he wanted to get his siesta, but he couldn't, because it was his duty to stay here and watch me. . . . And I felt some triumph just from that—because he couldn't leave me and get any shuteye at all because he had to be there. And I kept my arms up, jabbing him that way, never giving him the satisfaction. Then Slug was relieved by a new guard we called Scar-under-the-Eye. He was a short man, about four feet eight inches, with a large, grotesque scar under his right eye. He looked savage and he was. And now I couldn't keep my arms up there, so I relaxed, and he hit me, cuffing me around the head. Finally, after so much of that, I said I had to use the toilet bowl. I had to get relief from that banging around. I had been having bad diarrhea by then anyway, so they let me use it. I took the few minutes on that bowl to rest, and then Scar was on me again. I finally collapsed and fell over on the floor and he grabbed me by the hair and pulled me back upright with a jerk.

Now I felt I had nothing left to resist with. Nothing. I became irrational. I grabbed Scar and started yelling, "Bao cao!" all the time hanging on to him. He countered by shoving a dirty rag down my throat. I had had a blindfold on since he had started on me, tight across my nose, which cut off my breathing through the nasal passages. That rag in my throat now cut off my only other source of air. So I grabbed him again in desperation and I managed to shove him up against the wall, even in my leg irons, and immediately I let him go, sensing what I had done. He went tearing out of the room, called seven or eight of his buddies outside, and came charging back in. . . . They took the blindfold off, and when they looked at me I could tell they knew they had me. . . .

They put me back into the ropes, but this time they tossed another rope over the ceiling beam and ran it through a pulley. Next they pulled the rope and lifted me off the floor, so I was dangling a few feet off the cement. The pain of being hiked up was worse than any I had experienced up until then, all my weight being borne by my already beaten arms and shoulder sockets. Then Jawbone moved up to me and tried to tighten the ropes around my arms further. To do this, he had to drop me on the floor, and while I was standing there, glad for the reprieve from hanging on that pulley, he put his foot up against one of my arms so he could get leverage to tighten the ropes. As he did so I heard the bone crack, even before I felt it—and then the pain hit, stabbing a line of fire up into my brain. I shouted, panting heavily from the exertion of fighting those ropes, "You've broken my arm," addressing The Rat, who stood watching a few feet away. "No," he said simply, "we have not broken your arm. *You* have broken your arm." All the time WTG, who had come into the room, was trying to push that exposed bone back into place—a peculiar gesture, as if he were ashamed of its sticking out like that, or perhaps he was trying to ease it back in to save me the pain.

"We will punish step by step," The Rat went on. They would never say "torture." "Until you give us the information we want, we will punish." Then they would try variations like running ropes through my leg irons and pulling my legs in different ways. All the time, the same question. Who was the escape committee? And I kept hanging onto the same answer: "I don't know. We did not have an escape committee in the Zoo." Which was true. The escape was planned over in the Annex, in a nine-man room, but since I was the chief link in the communications network, they knew I had to know who was involved in that. I wondered if I should tell them something . . . anything to get them off me.

On the sixth night, I thought, "Surely tonight I will get some sleep." But they kept me even longer this time, making me kneel on that pitted cement with my arms up. Then it was back on that unbalanced stool, the slaps across my face now and then from The Rat with the tire sandal when I wouldn't talk. Sometimes there were long periods of nothing. Then I began to go into hallucinations. . . . The next morning WTG came in with the one cigarette, and I showed him the arm. He said, "Uh," and that was all. The point was that I was not going to get any medical aid until the interrogators were finished with me.

Thursday began. This time it was the ropes again, but now they put cords around my arms, tied wet damp cloths around the cords, and hooked these cords up to a battery. They gave me electric shock treatments on and off, and the pain was blinding, but mercifully I was so tired that it was only another blurring dimension of the pain I already had.

"You tell us now," the same words kept pounding at me from a long way off, "and we will stop." . . .

Yet I couldn't tell them anything. I still felt I should protect Guarino, Kopfman, and Connell. But why? I didn't know, except that I still entertained the idea that to tell was to open the gate on torture for them. . . . And so more electric shock. For three hours or so the treatment kept on. Once they stopped and put some food on the table, greens and soup, so I knew it had to be noon. The next time they gave me the same ration, it had to be four. I was still going through the ropes and the shock treatment, one jolting jab of voltage, jerking me in spasms, then another.

I felt myself sliding then. I was being beaten, whipped, falling to the point of nothingness. Death would be welcome. I wanted the pain to stop. . . . My will to hold on was dissolving.

Then I became aware that I wasn't going to come through this after all. I had drawn on all the reserves I had of pride, optimism, and the will to compete in this macabre game with my guards, the things I thought would carry me through. I had allowed hate for them to enter, for hatred fanned the flames of my determination to win. But I was miserably broken for all of that. I was bleeding, wracked with fever, my mind numbed by the electric shock, in and out of nightmarish hallucinations.

I thought of God then. In all this time, I had kept my mind on the pain that was coming in the next second, on bracing myself to outwit my guards, to dig in a little deeper to prevent myself from telling them anything. . . . Now I sensed I was losing for all of that, and I felt I was losing whatever I had of God as well. Was faith all a myth then? I had prayed, quoted Scripture, tried to live up to what I thought God would have of me—but what had I to show for it now? Where was that verse, the promise of Christ, that said, "Lo, I am with you always"? Where was the deliverance? *Where was He?* I had fought with all I had and lost—now was I going to lose the knowledge of God too?

I thought maybe I should try quoting more verses to remind God where I was, who I was, but none came. And I realized suddenly then that all the calling up of Scripture, all those repetitions of the Beatitudes, the Twenty-third Psalm, the prayers, maybe all of those had been meant to get God over to my side, to prevent the agony I was now in. That sense of the incompleteness of Christ in my life that I had felt even in the doing of all that, the big fragment of the unknown about God in my life, was it only now coming home? Would it only now dawn on me with new reality? I saw my life all over again in a few seconds—from that conversion experience in college all the way through to shootdown. In all that time, I had assumed Christ was in me, making of me a kind of model person in morality, good citizenship, love of country, family, and all the rest of it. But now I was struck with the fact that I had not entered into the sufferings of Christ in all that time. I had lived on the "good times" of Christianity, but I had never been tested by pain, as He had been, and the dimension missing in my life was tied directly to that.

In my befogged mind, and with the pulse of pain through me, I sensed that maybe God was trying to say something to me—that maybe there was something bigger, more real, more valuable than simply being the eternal optimist or the one who "gave the most to the game." Had I really given the most? Up to now, I knew I hadn't. It struck me then that God must have led me here, let me get shot down, that I might now enter into the totality of what it was all about to be in Him.

I tried to think it through while the pain kept up that mocking kind of rhythm in my body, telling me there was nothing to be gained from that kind of thinking either. But I remembered back in Heartbreak Hotel in Hanoi in 1967, at the height of my three-day torture, hearing church bells coming from somewhere in downtown Hanoi. I remembered hearing them at the very height of my pain and darkness then. I remembered then how it seemed that God was saying something to me in those soft bells, that he was not far away from me, that there was no pain or darkness so great that He would be outside it. I remembered what that had meant then, the hope it had given me, the renewed will to hang on.

There were no bells now, however, and I was worse off than I had been back there. I strained to hear those bells again, desperately wanting some evidence that God cared, that He was here with me in this place. I knew that if He didn't do something, reveal something of Himself to me, I could not make it. And, in my feeble way again, I said, "Lord . . . it's all Yours . . . whatever this means, whatever it is supposed to accomplish in me, whatever You have in mind now with all of this, it's Yours. . . ."

That was all I could say. That was all I had the mental strength to frame. I knew it wasn't much, but I meant it. It was the first time I had ever prayed so straight, so directly, so meaningfully. Whatever "commitment" I had given to Him up until then had never brought from me a prayer of surrender like that. I was totally willing now to accept whatever He had in mind.

It was a strange prayer for me, yet so absolutely right—even in my mind that could not fully focus on my words. But there wasn't a lot of time right then to dwell on it, because I was conscious of the guard there, waiting for the next slap . . . and yet there was something that preoccupied me even in that prayer, something that lifted the weight of fear from me. I didn't know what that prayer was supposed to accomplish for me. Nothing at all miraculous happened, and I wasn't really expecting anything.

But, in the next minute or so, I became aware of the fact that the ropes were being taken off my arms. The wet cords wrapped around my bare chest were removed. I remained there on my knees a long time, waiting, wondering, my eyes half shut. . . . And as I fell forward, too weak to stay kneeling, dropping down into my own blood and wastes, it suddenly seemed that the fifteen-watt bulb was turning on a glow of warmth within me. It was God. It had to be. I was alone, all the grim horrors of the past days and nights still with me, but now I had a moment of peace. I didn't know how to absorb the immensity of the moment, the sheer dimension of it, the mystery of it, yes—or the reality of it as well.

[McDaniel was given food and allowed to take a bath, but with his hands paralyzed from the torture, it was a difficult process.]

Friday came, and I faced it with a little more strength. God was here; that I was positive of now. I could face what was coming with a new sense of hope. But, almost as if they knew, the guards gave me the severest beating. I was beaten regularly by a two-man relay team with more than 120 licks with that fan belt on my bare buttocks. And after thirty licks or so, they would stop and put a watery, wintergreen-smelling substance on my buttocks, which had by now turned to hamburger. All of this, again, was to keep infection down. It was crazy—if I died in here, it seemed they wanted to be sure that no one would find any infection in my body. By now I was passing blood in my urine and through my anus, and that meant there were internal injuries. My eardrum had ruptured when they struck me across the head with my own shoe, and it too was oozing blood.

They continued to beat me that way until I thought I would go out of my mind with the pain. I said, "Okay, I'll tell you, stop." And they stopped. I took the few minutes while they waited just to get my breath and allow the pain to dissipate a little and then I said, "I don't have any answers."

So back to the beating. Then, knowing I couldn't go on anymore, that there just wasn't enough left in me to take it, I said, "Okay, I'll tell you." This time they hung over me, not allowing me time to fake it, and I gave them some names—but not the names they wanted me to confirm. I just told them the names of the senior officers in the building, and that I was the escape committee, nobody else. For some reason, they accepted that; and again, God must have been in it, because I did not confirm any of the information other tortured prisoners had given, and I had not given any names of the actual escape committee.

That Friday night I slept for the first time in a week. I was mistaken to think the interrogations were over or even the torture. But, as I slept, it was a sleep of assurance—God was not far outside this hell. If I had to go on with this nightmare, then I was sure He was with me. Nothing else mattered.

Excerpts from the 28 June 1995 Testimony of Carol Hrdlicka at the Hearing before the Military Personnel Subcommittee of the Committee on National Security

Carol Hrdlicka, like other wives of MIAs, impugns the U.S. government agencies involved in investigating the case of her husband Col. David L. Hrdlicka, who was known to be alive and in captivity at one time. His fate remains a mystery.

David L. Hrdlicka was thirty-three years old and a career Air Force pilot who loved his flying and believed in protecting his country. Our children were small—David M., seven years old, Denise, four years old, and Damian, three years old. He was looking forward to their growing up so they could share his love of the outdoors. He believed by protecting his country he was protecting his children and their future. He trusted the government. He would have never believed he would have been sent into a situation where his country would not come rescue him, or be denied by the very government he thought he was protecting.

David and I were married when I was just nineteen, so I matured as an Air Force wife. On May 18, 1965, I was only twenty-seven when the McConnell Air Force Base Commander, Col. Dannacher, told me the news David had been shot down. David was my whole life and I didn't expect to have the responsibility of raising three small children on my own. I always thought I would have David to help share ideas on how the children should be raised. I was also advised David had been lost in Laos, which was considered to be a secret war and I should not say anything because I could cause David to be killed. It was not until the 1990s that I discovered other families had also been told the same thing. I always did what was asked of me by casualty [Casualty Affairs] and trusted them to handle David's case because I believed they had the background and training in such matters, whereas I was just a military wife and didn't have the expertise. I believed them when they had a Presumptive Finding of Death Hearing in November 1972 and declared David "KIA." Only recently did I find out that it was an informal hearing and no evidence was presented at that time.

I followed their direction and tried to go on with my life. Thirteen years later, in 1990, I received a letter from DIA stating a source had information that David had tried to escape in 1989. That was a shock to me. Then DIA went on to debunk the source. In July of 1990 I received another report stating a "General Chaeng was suspected of holding D. Hrdlicka and friends." To my shock and horror, I began to realize I had been betrayed by the very government I had trusted for twenty-five years. I believe the only reason I received the Chaeng report was because Col. Millard Peck was in charge of the DIA POW/MIA office at the time. Col. Peck later resigned because of the way the families were denied the truth about their loved ones.

At that point I decided to start my own investigating. The more I questioned and asked for answers, the more I was ignored. I asked for evidence of David's dying in captivity and was told they had none. The DIA had used old reports, which they later admitted did not correlate to David, to falsify the fact he died. The JTF-FA were told by the Laotians David's grave had been blown up by the B-52 raids. Later the Laotians told them there was another grave site which was excavated. It contained no bone fragments or artifacts. Why do the Laotians keep changing their story?

This is the same thing that has been going on with the U.S. government for years. I will give you a couple of examples in my case of deceiving and lying by the DIA. In 1992, I sent a Freedom of Information request in on my husband, David L. Hrdlicka, asking for rescue attempts, all information, and information on an operation called "Duck Soup." Mr. Trowbridge of DIA, who coincidentally did the analysis on David's 1966 reports, responded—there was no such operation as "Duck Soup" and there had never been any rescue attempts for David. Low and behold, General Secord testified before the Senate Select Committee that there were rescue attempts for David made as late as 1967 and there were a raft of cables in CIA on that operation. Just in the last few months I have received documents describing there was such an operation as "Duck Soup," which took place in the Sam Neua area in June of 1965. . . . Guess whose name is in one of the cables regarding the Sam Neua area? David's.

The DIA continues to state I have been given all the information when my basic requests are, to this date, not answered. One outstanding example—I have requested the answer to a question being asked in a Stony Beach report where a U.S. investigator in June 1989 is questioning a General Chaeng "suspected of holding D. Hrdlicka and friends." I have asked for the "outcome of that conversation" for five years, and to date I have not received the answer. I have asked for David's authenticator code or whatever they used to identify him in case of his being shot down. I have not received that. In 1994, I received documents found by a private researcher, Mr. Roger Hall. These documents pertain to David by name and I had not been provided these documents by the very agencies that tell me they have given me "all the information." The Chairman, Senator John Kerry, promised, during my testimony before the Senate Select Committee, that he would get the answers on these same reports for me. Well, I have not heard a word from Senator John Kerry. Senator John Kerry promised, on the record, to arrange a meeting for me with the same Vietnamese officials he had met with. He has never fulfilled that promise either. I have offered many times, at my own expense, to meet with any officials and go over documents they say show David died in captivity. Yet, when I have asked for evidence that David died in captivity, I am told there is no evidence. The Department of Defense published a POW/MIA Factbook wherein it states there are intelligence reports that indicate David died in captivity. I have asked to see the exact reports they refer to, and am told I have the reports. Mr. Chairman, I do not have any report that has David's name on it stating he died in captivity. How is it that government agencies are able to circulate false information? Col. Schlatter of DIA has finally admitted there were reports correlated to David that do not correlate to David (many of which state a POW died on dates when David was still known to be alive). Where is Congressional oversight? Congressman Dornan, I beg you, get the Government Accounting Office to do an investigation of the DPMO, DIA, and the JTF-FA to include the shredding of documents in Bangkok. We have reports that show the JTF-FA wasting taxpayers' money and doing an unprofessional job closing our cases. They have committed fraud, waste, abuse. The taxpayers' money has been wasted under the guise of investigating cases of POW/MIAs. There are documents that show the DIA has known for years that remains of American servicemen are warehoused in Vietnam—why are we still putting young people at great risk and wasting $100 million each year of taxpayers' money digging at remote crash sites? Robert McNamara has just recently admitted he knowingly sent young men off to their deaths (to be murdered)—today we are sending young people into harm's way again by digging at remote crash sites. Don't these officials ever learn by their mistakes? It is high time that the DIA be held accountable for the waste and irresponsible handling of the POW issue through oversight. The DIA has been condemned by their own officials. In a

1985 memo form Thomas A. Brooks, Assistant Deputy Director for Collection Management to B. General Shuffelt, they are charged with shoddy work, not using basic analytical tools, not following up on cases, and all in all unprofessional. There were also condemned by General Tighe and again in 1991 by the resignation of Col. Peck. It is the responsibility of Congress to make sure the taxpayers' money is not wasted. At this point, it would be more cost effective to remove DIA from their current role and go to the private sector where honest, skilled investigators could be hired. I have exhausted every avenue, at my own expense, to find the truth about David's fate without the aid or assistance from the very agencies that were charged with solving his case. We even tried to take the evidence to the White House in November 1993. The President has not afforded the families the opportunity to present the evidence to him in person. The public is told there is much progress on the MIA issue but what the public is not told is, there were over 300 men left alive in captivity—what happened to them? These men were never missing; they were prisoners. However, there has been a systematic campaign to convince the American public that all the servicemen are MIA. That is not the case.

One prominent problem is getting the government agencies and officials to address the fact there were men captured alive, held in captivity, and never returned—what happened to them? These men were never in the missing status. Why is it then, the U.S. Government officials such as Assistant State Secretary Winston Lord, Deputy Assistant Secretary of Defense James Wold and Deputy Secretary for Veterans Affairs Hershel Gober will not address that question to the Vietnamese or Laotians? Instead, these officials consider the number one criteria remains and relay that to the Vietnamese and Laotians. On the recent high-level Presidential Delegation, the number one talking point still remains. The number one talking point should be, what happened to over 300 men that were alive and in the prison system? Did they just disappear into thin air? If these men died in captivity, then the Vietnamese and Laotians know where their remains are buried and should have returned them years ago. The Vietnamese were meticulous record keepers. Did they just lose these men in the prison system? It is a known fact that the Laotians admitted they held prisoners and would not release them until the U.S. officials came to Vientiane to make arrangements for the release of the prisoners they held. The Laotians said they would not release their prisoners through Hanoi. That meeting never took place and no one was released from Laos.

David was sent into harm's way by the U.S. government in a covert, unconstitutional war in Laos; where is oversight by Congress? David was an ordinary serviceman, so why was he used in a covert action? By being sent into a covert, unconstitutional war in Laos, there was no leverage to get him released? David's constitutional rights have been violated and I need the help of the Congress to protect David's rights. For many years I believed in and trusted every government official. I accepted as fact everything they told me about David's case. However, after seeing the evidence, I realize my trust has been betrayed. What is even worse, the U.S. government has betrayed their honorable servicemen. How is it that for more than 20 years this continual pattern of lying and deceiving the families has been allowed to continue? We have had many hearings and heard many promises; then in the end we are always patted on the head, and business as usual returns. . . . I've heard the lies and the promises; yet today, I am no closer to finding the truth about David's whereabouts or fate.

Source: Accounting 1996, 44–47.

Excerpts from the 28 June 1995 Statement of Judy Coady Rainey submitted to the Military Personnel Subcommittee of the Committee on National Security

Submitted for the record on behalf of Judy Coady Rainey by former POW Theodore Guy, this statement depicts the lack of sensitivity and dishonesty of which U.S. government employees were thought guilty in dealing with MIA families.

January 18, 1969, my brother, Capt. Robert Franklin Coady, went down in Laos. From January 1969 until 1974, the Air Force told my family that they had never received any information on my brother. In 1974 because of the lack of information, my sister-in-law had a Presumed Finding of

Death. The family did not object to the PFOD because the POWs had returned and we all knew that if the Air Force had received any information, we would have been told. So we thought my brother, who was promoted to major after he went down, had probably gone down with his plane. We had been told by the Air Force that if they ever received any information, we would be the first to know.

In 1991 I requested my brother's files from the Air Force. Two declassified reports were in the files: 1. A POW that was returned in August 1969 had the name "Bill Coady" (Bob went by the nickname of Bill as in Wild Bill). 2. A declassified source report was in the files dated 1972.

The POW Report. I got in touch with the POW office with the Air Force's help. The POW told me spelling was very important in the names he brought back. I tried to get the Air Force to send him his debriefing but was told debriefs were never released. DPMO (then DIA) told me that he spelled the name phonetically and that it was Cody or Cote. Mr. Desult at the Pentagon told me (after reviewing my brother's case for a trip that he planned to brief our family in Florida) that the name was spelled Coady on the original debriefing from Vientiane and all the other debriefings with the exception of an analyst debriefing at Andrews. . . .

When the Senate Select Committee was in progress, I wrote to them with all the problems I was having getting what I felt was all the information. My response from them was to go to the DIA and review my brother's files. My daughter and I went to Washington with the Senate Select Committee's letter in hand. Maj. Gittens from the DIA's office came down to talk with us. He said DIA didn't have any files. I said I had received files from DIA and couldn't we see them. He said they were classified files. I said couldn't he go through them and let us see what we could see and answer some questions for us. He said it was lunch time and that we could come back in two hours. We came back and he took us up to the DIA's office. He had four papers in hand, an FOIA [Freedom of Information Act request for information] from me, a FOIA from some other guy, a request from the Air Force for the files, and the partial debriefing from Andrews (that DIA sent to me). I started pulling out document after document that I had obtained, asking questions when he interrupted me and said, "I suppose you have the letter about the status change." My daughter jumped in and started to explain why her aunt had changed the status (my sister-in-law changing the status to PFOD had been thrown in our faces over and over again). Maj. Gittens said, "I'm not talking about that; I'm talking about the letter requesting a change from MIA to POW." I looked at him and said no. He said, "I can't talk about it; it's classified." He has since said he doesn't remember that and that we must have misunderstood. My daughter and I heard him very clearly. . . .

After I and the Senators had been told that I have all the information, I found out about the letter requesting a change in status from MIA to POW and also found out that they have crash site photos that were taken in March 1973. I keep finding out there is information I haven't known about. I requested to see the crash site photos and even had Senator [Richard] Shelby's (D–AL) office ask that I see them. I was told last year by Maj. Moore of DPMO that I could see them when I was in Washington last July. Only when I arrived in Washington, they could not find them. I was told that they would find them and my liaison office would bring them to me. When I asked Mr. Atkinson [George Atkinson, head of Casualty Affairs at Randolph Air Force Base] about them around January of this year, he thought that I had already seen them and he would check on them and get back to me. It's almost one year later, and I still don't have an answer, nor has Mr. Atkinson gotten back with me.

In 1993 I was told that JTF-FA was going to do a site survey on the crash site. I had been in touch with Senator Shelby's office trying to have family representation (my daughter and nephew) for the site survey. On September 1 I called CILHI and spoke with Johnny Webb. He told me that the survey was to be done in mid-October. Two weeks later I got a call from a colonel from Randolph [Air Force Base] telling me that a site survey had been done this past July and I would be getting the report from JTF-FA as soon as they got it. I can't tell you how upset I was that family wasn't allowed to be there. The report said that a Laotian that had seen the plane go down had taken them to the crash site. This Laotian had told them that the plane had gone down around 6 P.M. and it had been alone and that it was during monsoon season. This did not fit my brother's case. My brother's plane went down in January at 9:30 in the morning and there were all kinds of planes with him. So I started plotting all the planes that had gone down close to my brother who might have had some kind of prop planes because the source said that a prop had been taken away. Well, low and behold, Edward Leonard Jr. [another pilot] had gone down six months before my brother, and I believe it was eighteen miles from my brother's location. J.R. had walked out of that jungle and I thought he might be able to tell me something about what had happened to him and who (Vietnamese or Pathet Lao) were in that area. I called J.R. and was talking to him, telling him about the report and saying I didn't understand why they had related that crash site to my brother's. I told him what the Laotian had said about watching the plan go down at 6 P.M. and being all by itself. J.R. said, "Refresh my memory. When is monsoon season?" I said, "It is May through October." He said, "I guess it's my plane." He said he was by himself and it was dark and May when he went down.

I can't believe I have to do this kind of research when JTF-FA has all this information. JTF-FA still decided I was wrong and decided to do an excavation of the crash site. March of 1994 I was called by the mortuary and told that they had excavated all kinds of stuff, including an unrestated tooth. Well, again, I had already done my homework. I had talked to a forensic scientist and they had all told me you could not identify an unrestated tooth without DNA testing. The man from mortuary told me I didn't want DNA done because it would destroy the tooth and that we should see what the scientist there would say (maybe they could match it up with my brother's chart). I told him I still wanted DNA, that I didn't care if I had nothing to bury. I just wanted to know if the tooth belonged to my brother. I did hear from JTF-FA, saying that they didn't have anything to link the crash definitely to my brother's plane. The man from mortuary said I would hear from him in about three months about the tooth. Well—it's been 15 months and he hasn't called back! . . .

We have heard about family values in the last presidential election and now we are hearing about family values, bad movies, and music. We are worried about the youth of America. Congress has to set the example for these values. Congress might like to say that I am just an

emotional family member, but I didn't lie to my
government. My government lied to my family. The Air
Force tells me that the information I have is nothing and
proves nothing. Then why was it classified (if it wasn't
important) and why wasn't my family told about it? Why
was my brother put in a category 2, which is close to
category 1, which is a POW? It took twenty-two years to
find out my country had information on my brother. My
government did not come to my family to tell them. I had
to request the information. To me that is the height of
hypocrisy to tell my family that we would be the first to
know (twenty-two years later) of any information that
related to my brother.

Source: Accounting 1996, 22–26.

Excerpts from the Testimony of Patrick J. Cressman before the Military Personnel Subcommittee of the Committee on National Security, 28 June 1995 at the Hearing before the Military Personnel Subcommittee of the Committee on National Security

In the statement below, Patrick Cressman contests the assertion that his brother Peter R. Cressman was killed when his plane was shot down over Laos on 5 February 1973. He objects to the subsequent classification of KIA/BNR for all eight crew members of the aircraft, the use of a single tooth as proof of death, and the planning of a group burial. In the spring 1996 issue of the Department of Defense POW/MIA Newsletter, *Peter R. Cressman's case was listed as being resolved.*

Thank you for the opportunity to address the POW/MIA
situation, and more specifically, the loss incident of my
brother, Peter R. Cressman, Sergeant, United States Air
Force, and his seven crewmates, missing in action over
Laos, February 5, 1973. . . .

Despite the cease-fire, Peter's squadron was ordered
to continue with their highly classified airborne radio
direction finding operations, as part of a larger airborne
collection and reconnaissance program, which is utilized to
locate and monitor the communications and other signals
of foreign countries. Though Pete had concerns about the
squadron's activities being blatant violations of the Paris
Accords, he was told by the base legal office that he must

continue, in terms that made him abandon any thought of refusing to fly. A draft of a letter to his congressman, which detailed the various sections of the Accords which were being violated, was returned unfinished with his personal effects.

The flights that he and his crewmates flew were based out of several airfields in South Vietnam and Thailand. They flew in old C-47 "Gooney Bird" aircraft that had been refitted with a suite of radios and electronics gear, and redesignated as EC-47s. The EC-47s had been successfully employed for years in operations around the world during the Cold War and were used extensively during Vietnam. The "Electric Goons" were ideal for operations in Southeast Asia, where their reliable airframes could linger "low and slow" atop the jungle canopy, as the operators inside listened in on the enemy below, all the while looking like just another old C-47 cargo plane.

The specific aircraft involved here was an EC-47Q, callsign "BARON 52," which was based at Ubon Royal Thai Air Force Base (RTAFB) in eastern Thailand. It was one of several there assigned to a detachment of the 361st Tactical Electronic Warfare Squadron (TEWS). The EC-47Q was outfitted with oversized engines to compensate for the weight of the electronic gear, to make it more capable of flying on one engine in an emergency. BARON 52 had a crew of eight men—three pilots, primarily concerned with flying the aircraft; four electronic specialists, who operated the equipment in the rear of the plane; and a navigator, who was stationed in the rear of the aircraft, where he could both navigate the plane and assist the operators in plotting the locations of signal emitters. The crew consisted of Captain George Spitz, pilot and aircraft commander; Captain Arthur Bollinger, navigator; First Lieutenant Robert Bernhardt, first co-pilot; Second Lieutenant Severo Primm II, second co-pilot; Sergeant Joseph Matejov, airborne mission supervisor; Staff Sergeant Todd Melton; Sergeant Dale Brandenburg, and Sergeant Peter Cressman—systems operators. The four commissioned officers were members of the 361st TEWS, while the enlisted personnel were members of a detachment of the 6994th Security Squadron (USAFSS). These Security Service enlisted men all had to possess TOP SECRET/CRYPTO clearances, with access to Special Intelligence. Due to the nature of their work, the men of the Security Service were exposed to extremely sensitive information, not only concerning foreign operations, but to U.S. military operations and intelligence gathering activities as well. . . .

The assignment for the crew was, amongst other things, to locate, identify, and monitor the movements of North Vietnamese Army (NVA) armor and equipment moving along the Ho Chi Minh Trail in violation of the terms of the Paris Peace Accords. Of course, this developed into a potentially disastrous situation where we were violating the accords by watching the other side violate the accords. Both sides knew what was going on, but neither could confront the other without exposing their own infraction. At home, the Watergate scandal was starting to unravel, and with the cease-fire in place, as well as the pending release of POWs from North Vietnam, the Nixon administration could ill afford to risk a deterioration back to open hostilities. It was this environment that BARON 52 flew into on February 4, 1973, day eight of the cease-fire.

The mission had started out bad before the plane left the ground. The aircraft, whose tail numbers were 636, had mechanical problems so often that the squadron had nicknamed it "sick-three-sick," and that night was no exception. . . . After three hours of briefings and delays, BARON 52 departed from Ubon RTAFB towards its as-signed area (11G) in Laos at 11:05 P.M., arriving on station an hour later. . . . At 1:40 P.M., fifteen minutes after the first report of trouble, the EC-47 dropped from radar coverage. The crew of BARON 52 had joined the list of missing in action (MIA).

When the EC-47 failed to make its 2:00 A.M. scheduled call-in, MOONBEAN, two F-4 "OWL 5 & 6" fighter planes, and an AC-130 gunship, "SPECTRE 20," were diverted to the area of the plane's last reported position. BARON 62, another EC-47 working north of BARON 52, was contacted to perform a communications search. At 6:00 A.M., Search and Rescue (SAR) efforts officially began, with a "RUSTIC" OV-10 Forward Air Controller (FAC) aircraft being sent to visually search the area. All reported negative results.

At 8:24 A.M., the first of two intercepted North Vietnamese communications was reported indicating that a unit operating in southern Laos had captured four fliers shortly after the loss of BARON 52, in the same general area. These intercepts were not disclosed to the families for five years, and then only because Jack Anderson had told the Department of Defense that he was going to air it. These two messages were also withheld from the Wing Commander, who ultimately declared all eight men dead. This information becomes critical in the case due to the fact that no multi-crew aircraft were lost in this area since December of 1972, nor subsequent to the crash, leaving the back-end crew from BARON 52 the most likely candidates to be Americans, wandering around that area of southern Laos before dawn, wearing flying attire. This may also be supported by a report, apparently still classified, specifically referred to at least twice by independent sources, of a "friendly" Lao road watcher who saw the four "clean shaven and in their flightsuits" being led away under NVA guard. One Security Service airman has reported, to date without confirmation, that he attended a briefing a day or two after the loss of BARON 52 in which he recalls the briefing officer stating that the commander of the EC-47 had radioed that the back-enders were out and had good parachutes. This account is still under investigation.

The wreckage was finally located on February 7 through the use of photographic reconnaissance flights of the area. On the morning of February 9, a search team was launched from Nakom Phanom RTAFB to inspect the wreckage. A ground team of three USAF pararescue specialists (P.J.s) and Msgt. Ronald Schofield of the 6994th Security Squadron, were lowered into the crashsite, while Sgt. Keen (also of the 6994th) and a photographer stayed on board the helicopter. Another helicopter was holding in an area five miles away in case they were needed. The team found the wreck of BARON 52 inverted, intact from nose to tail, with the wingtips broken off and located some 400 meters up the ridge. The crashsite was obviously the site of a fire after it came to rest due to the burned wreckage and foliage. The team found three bodies; the pilot (Spitz), the first co-pilot (Primm), and the second co-pilot (Bernhardt) all at their duty stations. One of the searchers said that he saw "one other possible—repeat—possible body" in the craft, though none of the others on the scene did. The remains of Lt. Bernhardt were the only ones that Schofield and one of the P.J.s could reach, and after attempts to pull them free [he] tore them in half, [for] only partial remains were recoverable. (Bernhardt's remains were subsequently confirmed, and his status changed from MIA to Killed in Action. All the other crewmembers were maintained in their MIA status.) In the rear, where the navigator and enlisted men were stationed, neither they, their parachutes and vests, nor their electronics gear was found. The emergency exit "kick-out" door was not at the site. These findings are very significant since it was the standard operating procedure at the time to either destroy the equipment with a fire ax and/or jettison it so that it would be destroyed and strewn along the ground, preventing it from falling into enemy hands intact. Another important fact is that the fuselage of the aircraft had melted down into about an 18 to 24 inch crawlspace after impact, which would not be possible if the large electronic consoles were still in place providing structural support. All these indications support the conclusion that the crew had ample time to react to an emergency (15 minutes from reported AAA [anti-aircraft artillery] fire to last contact), by throwing the equipment, and then themselves out of the craft, leaving the highly experienced Spitz and the others to try and successfully return to a friendly base. Though the intent of the flight crew is admittedly conjecture, the absence of all non-essential personnel and the locations in which bodies were found in the wreck strongly support that conclusion.

Another intercept was reported on February 17, stating, in pertinent part, that "the people involved in the South Laotian Campaign have shot down one aircraft and captured the pilots." Once again, no other U.S. aircraft were lost in that area during the same period.

On February 22, Col. Francis Humphries, without the benefit of any new evidence, declared the entire crew Killed in Action (KIA). This despite the objections of his own intelligence staff, disregarding any signal intelligence he may have seen, and in complete disregard of the fact that only three of the crew were in the crash to die from it. In defense of the colonel's action, there is considerable evidence indicating that this wrongful, and possibly illegal action, was a decision made at the higher levels of government. . . .

On February 23, 1987, DIA's Special Office for

POW/MIA completed their "exhaustive" analysis of all
available intelligence and operational data surrounding the
EC-47Q loss. It was, unfortunately, focused on debunking
the collection and analysis of virtually every piece of work
done on this case up to that date. In one statement,
prepared by one man, all the intercepted information, and
all the analysis by the real-time hands-on collectors and
analysts, no longer related in any way to BARON 52. The
ground search in 1973 suddenly found four rather than
three bodies in the wreck, and had "conclusively"
determined the fate of all eight crewmembers. In fact, none
of the ground team members ever tried to say that they had
accounted for all eight crewmembers. In fact, none of the
ground team members ever tried to say that they had
accounted for all eight men at the crashsite. Each and every
conclusion in that evaluation does not hold up to the least
bit of scrutiny. The report is full of holes, misrepresenta-
tions, and deception. And Destatte himself admitted in
1992 that he had never interviewed the analysts who had
collected the intercepts nor the analysts who, with all the
current information available to them, evaluated the
information and prepared their comments and conclusions.
What kind of re-evaluation doesn't include asking someone
what they said and why? It was plain to see exactly what
kind of "re-evaluation" had been done.

On February 9, 1994, Mr. Garold Huey of Air Force
Mortuary Affairs came to Florida to present the families of
Sergeants Cressman and Matejov with the results of an
excavation allegedly performed at the crashsite of BARON
52. . . . We were informed that the site had been thoroughly
examined over several days, and that what was found
confirmed the fate of all eight crewmembers. Remarkably,
in this thorough investigation, NO remains were found belonging to the three bodies we know were in the wreckage
in 1973. In fact, the only "remains" found of these eight
men were a handful of bone fragments which the Central
Identification Lab admits cannot even be proven to be
human. As well as a piece of a single tooth which they
equate to my brother, Sgt. Peter Cressman. . . .

Based upon the x-rays of this single piece of tooth,
which remains the only item with potential identification
value, we were advised that the Department of Defense
considers this material to be the commingled remains of all

eight fliers, and further that they were planning a "Group Burial" of the crew. When questioned if this burial would take place despite the objections of our two families, Mr. Huey replied, "Yes." . . . We have repeatedly asked for DNA confirmation of this alleged "ID," at our own expense, only to be told that this will not be permitted because DNA testing is a destructive procedure. . . .

The declaration of death in this case cannot be defended; it was wrong in 1973, and it remains wrong today. Virtually all evidence shows that the back-enders had bailed out avoiding the crash and were captured. No amount of back peddling or "re- evaluation" can change that.

Source: Accounting 1996, 51–56.

Excerpts from the Testimony of Colonel Theodore Guy

Now-retired U.S. Air Force Colonel Theodore Guy, a fighter pilot during the Vietnam war, was shot down over Laos in 1968, where he was captured by North Vietnamese troops. Guy was held prisoner in several different Hanoi area prisons and released during Operation Homecoming in 1973. In his 1995 statement to the Committee on National Security, excerpted below, Guy explains why he believes that American prisoners of war could still be alive.

[Guy introduces himself to the committee and cites his military record.] When I was downed, I was flying an interdiction mission in southern Laos along highway 5, which runs from Khesan to Tchepone. The target was an automatic weapons position overlooking highway 5. On my third bomb pass, my aircraft, an F-4C, was rocked by a violent explosion. We later determined after the war that the explosion was caused by one of my 750-pound high drags going slick and detonating approximately 50 feet below the aircraft. The aircraft was severely damaged, but we managed to get to 12,000 feet, where we lost control. I told my back seater, Major Don Lyon, to eject, which he acknowledged. The next thing I remember is floating down in my parachute. I am certain the aircraft blew up and that Major Lyon never got out of the aircraft.

I arrived on the ground several minutes later, and after a short fire fight with the Vietnamese, I was captured.

During this gun battle—which I lost—I was wounded.

After seizure I was dragged and carried about a half mile until we came to what appeared to be a staging area. I estimate that there were approximately two battalions of 1,000 troops in this area. All wore, what appeared to be, fairly fresh green summer clothing and white tennis shoes. As I was dragged into this area, I noticed numerous wires running off into the jungle. It consisted of red and green strands. One of the strands was hooked to a field telephone, which was in use.

On 26 March I departed this area by jeep on my way North. We spent the first night in Tchepone, and there was no doubt in my mind that we were expected when we got there. The next day we started up the Ho Chi Minh Trail.

Travel was by jeep and truck with several long walks. I am convinced that all the personnel we saw along the trail were Vietnamese [rather than Laotian]. We spent one 24-hour period in a Laotian village, and again there was no doubt who was in charge—the Vietnamese. I spent eight days on the trail. Until we were well North, about half of our travel was at night. We would usually stop traveling an hour before sunrise. The majority of the time we would pause at caves located a short distance from the trail. While traveling North, I continually observed the same type of communications wire running along the road. Every place we stopped had a telephone that appeared to be hooked up to this wire. It seemed to me that I was expected at every stop. On two occasions I was addressed by my Vietnamese name of Gee, before my traveling guards had a chance to converse.

Based on the above information, I have concluded the following: From what I observed, there is no doubt in my mind that Laos was wired and wired well. My movements were obviously continually reported. Communication between check-points was excellent. Based on my observations, I believe all trail activities were closely monitored—probably from Hanoi.

I arrived in Vinh, North Vietnam, on the 3rd of April, where I had my first real interrogation. These guys did not fool around and managed to dislocate my right shoulder. I was told that the U.S. bombing had stopped as of 31 March. I was held in Vinh for three days in a compound that was heavily reinforced and appeared to be a prison with several cells. I saw no other Americans. On 7 April we departed Vinh by jeep and arrived in Hanoi in the evening.

I spent the next three days at the infamous Hoa Lo prison "Hanoi Hilton" where I learned that the Vietnamese knew more about Camh Rhan Air Force Base and the 12 TFW [Tactical Fighter Wing] than I did and their information was as current as mine. I was continually asked where I was captured. During the evening of 10 April, I was transferred to a camp known as the "Plantation Gardens," where I remained until 17 December 1969.

During my many early interrogations at the "Plantation," the Vietnamese tried to convince me I was captured in North Vietnam. I insisted that I was captured in Laos. During the fall of 1968, it became evident that I was the only known POW captured in Laos. I then changed my story and "admitted" that I was really captured in South Vietnam but could have drifted in my parachute to North Vietnam. Ernie Brace, CIA, captured in Laos, arrived at the "Plantation" in October 1969, so there were now two of us. However, I maintained my South Vietnam story, mostly to hide my knowledge that another Lao POW had arrived at the "Plantation." During one of my interrogations in the "Hilton" in May 1970, "The Bug" informed me that they had rechecked their records and I had indeed been captured in South Vietnam. Shortly thereafter, I was transferred to the hell hole known as "Farnsworth," that contained only South Vietnam captives.

Based upon the above information, I have concluded the following: The Vietnamese kept detailed records on all POWs, and summaries of these records followed the POW from camp to camp.

The night after the Sontay raid, all POWs from "Camp Farnsworth" were transferred back to Hanoi and the camp known as "The Plantation." I was returned to the same cell—still in solitary—that I had occupied from April 1968 to December 1969.

The widely reported change in treatment towards the POWs that occurred after Ho Chi Minh's death in 1969 did not occur in our camp (e.g., those of us captured in Laos and South Vietnam). Harsh treatment, near starvation diet, isolation, and beatings remained in effect until the summer of 1972. Tolerable conditions prevailed after the resumption of the bombing of North Vietnam and the mining of Hyphong Harbor. In July 1971 six others captured in Laos were transferred from "The Hilton" to the "Plantation"; among them was Ernie Brace. They were amazed at our treatment and informed me that it was much better in other camps.

It appeared to me that the Vietnamese were systematically grouping their POWs. All those captured in the North were at the "Hilton," while those captured in Laos and South Vietnam were at the "Plantation." There were no departures from the "Plantation" to other camps, only incoming Laos and South Vietnam POWs.

Because the North Vietnamese continually denied any association with Laos or South Vietnam, other than providing support, I came to the conclusion that we were not going to be released at the end of hostilities. The word was passed to my command that we were to prepare for the long haul—which I felt could be as long as 20 years. The majority of the POWs accepted this with a fighting spirit which made me extremely proud.

In December 1972, on the third day of the B-52 raids, all of us were transferred to the "Hilton" and housed in the area known as "Little Las Vegas." The ten other Lao captives were kept separate from myself, who by this time was considered a regular South Vietnam captive. We managed to establish contact with the 4th Allied POW Wing. All names were passed, and I passed my fears to the 4th commander, John Flynn, that I felt we might not be released.

All of us were released, although considerable wheeling and dealing was necessary to gain the release of the ten remaining Lao captives.

Based upon the above information, I have concluded the following: I believe there is a good possibility that if the December bombing had not occurred and we had stayed at the "Plantation," we would not have been released during Operation Homecoming. I base this on the fact that our captors denied that they had troops in the South or Laos. Also our treatment remained far worse than those captured in the North.

I retired from the Air Force in 1975. From the time of my release until mid-1991, the thought that any POWs were left behind never crossed my mind. In fact, I spoke to hundreds of MIA families and tens of thousands of people about my POW experiences. My message was always the same. All the POWs are home. There are no MIAs—they are all dead. All that wanted to come home are home. I told the families to forget their sons, fathers, uncles, etc. and to get on with their lives. After I explained that all the POWs that were captured ended up in Hanoi, and that all the names were known, the majority of families accepted their fate.

At a POW Dining In at Randolph Air Force Base in March 1991, I had a long discussion with Brigadier General Robinson Risner (POW Sept. 1966 to Feb. 1973) and Lieutenant General John P. Flynn (POW Oct. 1967 to March 1973). General Flynn was the commander of the 4th Allied POW Wing. Both were members of the Tighe Commission headed by Lieutenant General Eugene Tighe. Both Robbie and John firmly believed that American POWs were left behind and that there was a good possibility that some were still alive. They based their beliefs that our country left men behind on information learned from their involvement with the Tighe Commission. I could not believe that the United States would knowingly abandon any of her fighting men. The very thought of this was repulsive and unacceptable.

In June 1991 I was called by a Mr. George Atkinson, Casualty Affairs, MPC, Randolph Air Force Base, Texas. Mr. Atkinson asked me if I would come out to Randolph and talk to a young lady whose brother was shot down in Laos in 1967. Mr. Atkinson was well aware of the fact that I felt very strongly that all POWs were home that were coming home.

The next morning I spent several hours with the MIA sister. At lunch I repeated all my theories about the missing POWs and MIAs. The sister agreed that it was time to get on with her life and put her brother behind. However, she did request that I review her brother's file prior to returning home. I did and I was shocked.

The brother was a back seat navigator on a B-57 bombing sortie in northern Laos in the fall of 1967. His aircraft did not return and he was listed as MIA. One year later his status was changed to PFOD [presumptive finding of death] with no objections from his wife. The Air Force never notified the blood family members, assuming the wife would take care of the matter. Because of great family difficulties, this never happened. After hearing nothing from the sister-in-law for years, the sister contacted Randolph about her brother.

The folder contained many references to the brother. There were refugee sighting reports and several identifications by Lao and Vietnamese refugees who had gotten out of Vietnam and Laos. I took the folder to the head of the Casualty Affairs Branch. My comments were, "How can this type of information be in this folder? He was shot down in 1967 and declared KIA a year later. Either he is dead or he isn't. Why would information continue to flow to the folder if the man was dead? Someone has to be either very stupid and thought no one would notice, or this man is alive and no one gives a damn!" Some of the sightings were in the late 1970s and mid-1980s! I concluded by saying, "No wonder the families do not believe what they are told.". . .

After the meeting with Robbie and John and this meeting at the Casualty Affairs Branch, doubts began to creep into my mind. I started reading [about] and contacting as many MIA/POW families as I could. I corresponded with other activists and talked to many Vietnam veterans. There was no hesitation from the people I talked to. Men were left behind, and worse; there had been little to no attempt to account for anyone that disappeared in Laos. The deeper I dug, the more convinced I became that men were abandoned and that there was a good possibility that some were still alive. Watching the Select Committee on television and reading the *Select Committee on POW/MIA Affairs, Senate Report 103-1* further convinced me that much was being hidden and withheld about the POW issue. [Guy, citing statistics, indicated that of the servicemen missing in Laos, only 1.9 percent were returned at Operation Homecoming. Of the total servicemen missing in North Vietnam, 36.8 percent were returned during Operation Homecoming.]

Based on the above information, I have concluded the following: All 11 of us that were captured in Laos have one thing in common. We were all captured by regular North Vietnamese troops. Initially, I believed that because we were captured by North Vietnamese was the sole reason we were released. However, I now feel this had little bearing on our release. Based on the large number of NVA regulars I observed in Laos, I submit that many, many more were captured by Vietnamese forces. Many people do not think the Vietnamese are very knowledgeable. I personally believe the ones that had control over the POWs were brilliant. I also believe they foresaw the possibility that we would prove that their forces were dominant in Laos. We (the 11) were the tokens that were in their long range plans to be released—if pressured. OR THEY MADE A HUGE MISTAKE!

It has often been stated, both unofficially and officially, that if there were men abandoned and left behind after the Vietnam war, they could not survive very long under the harsh conditions. I disagree. I am convinced that with high morale and determination, the American fighting man can survive indefinitely; even under the most austere atmosphere. To support this, one only has to review two recent events.

The case of the F-16 piloted by Captain Scott O'Grady that was shot down on June 2nd [during the Persian Gulf War] has many similarities between aircraft lost in Laos during the Vietnam war. O'Grady's wingman, Captain Bob Wright, never saw O'Grady eject or a parachute nor was any contact established between the two once O'Grady landed. Yet, Captain O'Grady was rescued six days later. If he had been captured, would he now be carried as MIA? In Laos, there were many ejections which were observed by wingmen. A parachute *was* seen and contact *was* established with the downed pilot. Only in Laos, he disappeared.

On April 14, 1975, the *New York Times* reported that hundreds of Vietnamese, who were employed by the Central Intelligence Agency (CIA) and military, were captured and imprisoned in the mid-1960s. The U.S. government wrote them off; however, in the late 1980s the survivors were released. Sixty-four have applied for refugee status under the Orderly Departure Program. The Immigration and Naturalization Service denied admission. Are not the survivors living proof that man can survive long periods of internment under very harsh conditions?

Based on the above information, I have concluded the following: It is possible to survive for long periods of time under the most severe conditions. I am convinced that the majority of my command could still be alive today if we had not been released.

Source: Accounting 1996, 9–15.

Speeches and Quotes

Position on United States Relations with Vietnam in the Context of POW/MIA Progress, 26 January 1994

This excerpt from a memorandum by Ann Mills Griffiths, executive director of the National League of Families of American Prisoners and Missing in Southeast Asia, was submitted to Congress by Senator Bob Dole on 26 January 1994, shortly before the lifting of the trade embargo against Vietnam by President Clinton. In her statement, Griffiths expressed reservations about lifting the trade embargo because she believes that the Vietnamese are withholding the remains of U.S. servicemen and archival documents regarding their fate. Griffiths supported the Dole/Smith amendment to the Kerry/McCain amendment to S. 1281 of the State Department Authorization Bill. The Kerry/McCain amendment urged that the embargo be lifted immediately, while the Smith/Dole amendment would have required the president to determine that Vietnam had provided the remains and information that the U.S. government believes it to possess prior to lifting the embargo. The Kerry/McCain amendment passed. The Dole/Smith amendment did not.

If Vietnam unilaterally provides the remains of Americans and incident-related documents which the U.S. intelligence community believes they are withholding, the National League of POW/MIA Families is not opposed to reciprocal steps by the United States to improve diplomatic and economic relations. We have supported that approach since 1989 and advocated humanitarian assistance since 1986. What we oppose are steps by the United States to meet Vietnam's economic and political objectives before their leadership authorizes unilateral actions which would rapidly account for hundreds of Americans.

Our position on living POWs is that Americans were alive at the end of the war, have not been returned, must be assumed still alive without evidence to the contrary, and that the Government of Vietnam can easily resolve these questions. If Americans last known alive in captivity are no longer living, their remains should be readily available to Vietnamese authorities. Field searches are not necessary to resolve these cases; a political decision by the Vietnamese leadership is required.

Vietnam's Ability to Rapidly Account for Missing Americans

Family members, veterans and other League supporters throughout the country oppose further steps to lift the U.S. embargo or improve political relations until Hanoi makes the decision to cooperate fully and stops manipulating this issue. The League supports reciprocity, but not when Vietnam is clearly withholding the answers from families.

One way of viewing what the United States knows and what Vietnam can do is by looking at what Hanoi has not, but could have, done. U.S. intelligence and other data confirms over 200 unaccounted for discrepancy cases of Americans last known alive, reported alive, or in close proximity to capture. In approximately 100 of these cases, investigations have reportedly been sufficient to confirm death. Hanoi knows that these are highest priority cases, as they relate directly to the live prisoner issue. If deceased, remains of these Americans are logically the most readily available for repatriation since they were captured on the ground or in direct proximity to PAVN [People's Army of Vietnam] forces. Yet, Vietnam has purposely avoided accounting for these Americans, allowing only investigations to determine fate, while signaling availability of more data.

U.S. wartime and post-war reporting on specific cases, captured Vietnamese documents concerning the handling of U.S. prisoners and casualties, and debriefs of Communist Vietnamese captives, reinforced by U.S. monitored directives and other reporting, formed a clear picture of a comprehensive North Vietnamese system for collection of remains and information dating back to the

French-Indochina War. Specific sources such as the mortician in 1979, substantiated by others in the 1980s, highlighted remains storage as a key factor in obtaining accountability.

During the war and since, the Vietnamese Communists placed great value on the recovery and/or recording of burial locations of U.S. remains. During the war, if jeopardized by imminent discovery or recovery by U.S. forces, burial was immediate to hide the remains, then disinterment when possible, photography and reburial, or transfer to Hanoi if feasible. Evidence of this process is confirmed by U.S. intelligence.

Assessment of community-wide intelligence serves as the basis for U.S. expectations that hundreds of Americans could be rapidly accounted for with unilateral Vietnamese action to repatriate remains. In 1986–1987, the entire intelligence community maintained higher estimates, but the numbers were subsequently further screened to establish the most realistic targets for the Vietnamese government to meet.

Forensic evidence serves as another basis for establishing expectations. Roughly 65% of the 279 identified remains returned from Vietnam since the end of the war have shown evidence of both above and below ground storage. This is hard evidence, confirmed by forensic scientists.

After two years of no results from the Vietnamese in 1979–1980, during a September 1982 ABC "Nightline" program, SRV Foreign Minister Nguyen Co Thach flatly denied holding any U.S. remains, as had SRV officials throughout the Carter Administration; Vietnam returned 8 stored remains in 1983. Negotiations for a two-year plan in 1985 brought the largest number of remains obtained to that point; nearly all 38 showed evidence of storage. In 1987, negotiations resulted in the largest number of remains returned during one year, 62 in 1988, 30 of which were returned at one time. Nearly all were virtually complete skeletons which showed clear evidence of storage; there are other more recent examples.

The total number of identified remains returned from Vietnam with evidence of storage does not equal the number reported stored by valid sources, nor come close to the U.S. government assessment of remains available for unilateral SRV repatriation. Evidence of storage exists on remains returned this year, but not yet identified; an important signal was also sent by the SRV in a 1989 stored-remains repatriation. Both instances revealed province-level storage/curation; there are many other examples.

Vietnamese officials have also admitted storage of remains. In 1985, following up an initiative through a regional government, a National Security Council (NSC) official met privately with a politburo-level Vietnamese official during a NSC-led U.S. delegation to Hanoi. The carefully drawn plan was for negotiations on live prisoners and remains. The SRV foreign minister indicated that no live prisoners were on the table for discussion, but that the hundreds of remains discussed through the third party were.

In order to test the scope of Vietnamese knowledge, two specific cases were officially presented to SRV officials in 1985–1986 with a request for their unilateral assistance; both losses occurred in Lao territory under PAVN control during the war. One was returned unilaterally in 1988, 98% complete and stored above ground since the incident. Vietnam has unilaterally repatriated stored remains from remote locations spanning the entire war.

There is continuity today. In 1991 and 1993, the SRV provided graves registration lists with names of unaccounted for Americans. Inclusion of these names was likely again purposeful, as was filtering through private channels photographs of dead, unaccounted for Americans whose remains have not yet been returned. Combat photography was directed by the DRV/SRV government; DRV/PRG (Provisional Revolutionary Government of South Vietnam) soldiers did not own personal cameras, much less carry them. Regardless of mixed or conflicting signals on both sides, these and other actions by SRV officials are intended to signal the United States of remains availability.

Information obtained from field operations after the war, including recent Joint Task Force–Full Accounting (JTF-FA) activities, also reveals that central DRV/SRV authorities systematically recovered American remains. Eyewitnesses reported central authorities arriving to supervise remains recoveries of Americans not yet accounted for. As long as Vietnam continues to benefit financially and politically from field investigations of these same cases, Hanoi has little motivation to unilaterally repatriate remains now being withheld.

Source: Griffiths, 1994, S211.

Winston Lord's Speech on Lifting the Trade Embargo, 9 February 1994

Winston Lord, assistant secretary for East Asian and Pacific Affairs, spoke before the Senate Foreign Relations Subcommittee on East Asian and Pacific Affairs on 9 February 1994, a week before President Clinton lifted the trade embargo against the Socialist Republic of Vietnam. Attempting to justify President Clinton's lifting of the trade embargo, Lord discussed four areas of effort by the Clinton administration: remains, discrepancy cases, trilateral cooperation with Laos and Vietnam, and archival research.

Mr. Chairman, distinguished members of the committee: The investigation of Case 0954 began in October 1992 when local Vietnamese villagers unilaterally returned 531 bone fragments, 16 teeth, an ID tag, a Geneva Convention Card, and an aircraft data plate to local officials during the 20th Joint Field Activity. A CILHI team climbed to the site in November 1992, conducted a site survey, and recommended against excavation due to the hazards involved in climbing to the site and the difficulty of the terrain. The Commander of the Joint Task Force–Full Accounting directed that his detachment commander in Hanoi, an experienced infantry officer, go to the site and determine whether an excavation could be done safely. In March 1993, the detachment commander and another detachment member traveled to the remote site. Three aerial reconnaissance attempts failed to locate a landing zone close to the site due to the ruggedness of the terrain. From the nearest road, the team climbed uphill for five hours to a small farm inhabited by only two people, remained overnight, and the next day climbed an additional two hours to reach the site. The site was located at an elevation of 4,780 feet on the side of a mountainous rock formation that varies in slope from 30 to 60 degrees. The detachment commander determined that an excavation could be done safely, but it would be extremely difficult and would require a hand-picked team in top physical shape. Prior to the 26th Joint Field Activity, the Vietnamese cut a helicopter landing zone suitable for an MI-8 on the side of the mountain, thereby reducing the climbing time to the site.

Over a two-day period, six MI-8 sorties transported the 12 U.S. and 15 Vietnamese recovery team members with their water, equipment, and supplies to the landing zone. From the landing zone the team carried equipment for about two hours over extremely rugged terrain to a base camp. The crash site was over an hour's climb from the base camp and the terrain was so steep that at points it required scaling rock faces hand over hand. Over the next two-and-a-half weeks, the team climbed an hour each day from the base camp to the site, excavated at the site, then climbed for an hour back to the base camp.

The immediate area of the crash is a rocky slope 40 to 45 degrees in grade. Working from the lowest elevation to the heights at the site, the team worked over the next 16 days removing surface rock and scraping and sifting through screens the associated soil, aircraft debris, and human remains. The excavation resulted in 187 bone fragments, 16 human teeth, personal effects, life support equipment, and other wreckage. This excavation, along with the earlier unilateral turn-in, resulted in a total of 718 bone fragments and 16 teeth.

Mr. Chairman, that is the story of just one case among the hundreds that brave and devoted Americans are pursuing every day—in the jungles and on the mountains of Vietnam, Laos, and Cambodia; in the laboratories in Hawaii; in Pentagon offices; and in hearts and minds. This brief vignette illustrates not only the labors of Americans but also the intensified cooperation of the Vietnamese. And it shows we are getting results from a process that is painstaking, incremental, and will last for decades.

Against this backdrop, I welcome the opportunity to appear before you to discuss President Clinton's decisions last week to lift the trade embargo against Vietnam and to establish a liaison office in Hanoi.

The President took these steps because he was convinced that they offered the best way to achieve the fullest possible accounting for our POW/MIAs. At the outset, I want to emphasize that his decisions were based on that single judgment. Of course the Administration is not oblivious to the potential economic and geo-political benefits that may now begin to unfold. But such benefits would flow from last week's decisions; they were not the reasons for them.

Thus, as the President stated, the POW/MIA issue will remain a central focus of our relationship with Vietnam. We will continue to require, in his words, "more progress, more cooperation, and more answers."

The Administration's Search for Answers

President Clinton's decisions were preceded by an intense government-wide effort during the first year of his term. This Administration has devoted more resources to the POW/MIA accounting effort than any previous one; there are now more than 500 military and civilian personnel assigned to this task under the leadership of Secretary Perry, General Shalikashvili, Chairman of the Joint Chiefs of Staff and the Commander in Chief of the Pacific, Admiral Larson.

From the beginning, President Clinton has worked hard to change the way the government handles information about the POW/MIA issue to ensure full disclosure. On Memorial Day, he pledged to declassify and make available all possible government documents related to our unaccounted-for men. On Veterans Day, we fulfilled that pledge. The State Department reviewed about 200,000 pages of documents, and we declassified and released more than 99%. The small amount of material that has been withheld from release consists of matters relating to personal privacy or sensitive foreign policy discussions. The public can gain access to the released documents at our Freedom of Information reading room. I understand that the Defense Department declassified about 1.5 million pages of documents, which are available at the Library of Congress. President Clinton and his top advisers have also made extraordinary efforts to consult many groups that share his concern for the POW/MIA issue. He insisted that all points of view be carefully considered. As is well known, some of those we consulted do not support lifting the embargo at this time.

This Administration has provided American veterans

organizations an unprecedented role on this issue. For the first time, leaders of major groups accompanied a presidential delegation to Vietnam last July to press for more progress. We have continued to meet with those organizations and other representatives of veterans. The various leaders and their constituents hold diverse perspectives, and we have benefitted from them all.

We have also consulted regularly with the National League of Families of POWs and MIAs [National League of Families of American Prisoners and Missing in Southeast Asia]. I would like to pay tribute to that organization, which, during the 1980s, was instrumental in pushing our government to do more to account for our missing men. Much of the credit is due to Mrs. Ann Mills Griffiths, the League's executive director and the sister of one of our missing. We invited the League to join the July mission to Vietnam, but they were unable to participate.

To the veterans and families, let me repeat that this Administration remains steadfast in its determination to achieve the fullest possible accounting. Our doors remain open. We encourage them to continue working with us toward our common goal. As the President stated last week, this spring he will send another high-level delegation to Vietnam, and will, again, invite the veterans organizations and the League of Families to participate.

The President and his advisers also sought the views of a large, bipartisan group of senators and representatives, including members of this sub-committee and many who were, themselves, prisoners of war or served in Vietnam, including the chairman of this subcommittee [Senator John Kerry].

Finally, the President has relied heavily on the information and advice provided by his military and civilian advisers here and on the ground.

Chronology of Developments

These intensive consultations are part of the careful, steady course on Vietnam that the President has charted during the first year of his Administration. Let me briefly review the events that led to the President's decision to end the embargo and establish reciprocal liaison offices.

The first milestone was the April 1993 mission to

Hanoi of Gen. John Vessey. Mr. Chairman, I would like to use this opportunity to salute Jack Vessey. Much of what we have accomplished on the POW/MIA issue is due to the dedicated labors of this patriot who has served three Presidents as Special Emissary to Hanoi. Entering the army as a private, he rose to the highest position in our armed forces—Chairman of the Joint Chiefs of Staff. He gave 46 years of outstanding service to the nation. A grateful country could not have asked for more, but General Vessey had more to offer. He devoted himself to seeking the answers to the questions that have plagued the families of the missing. In 1987, he went on the first of six missions to Hanoi as Special Emissary. His work led to the establishment of the Joint Task Force–Full Accounting, and to our full-time POW/MIA office in Hanoi.

During General Vessey's April mission we were able to investigate the information we had just received from the archives of the former Soviet Union. Hanoi agreed to establish special teams to investigate the remaining discrepancy cases. For the first time, we received documents from Vietnam's wartime general political directorate.

On July 2 last year, the President announced two new steps toward our goal of the fullest possible accounting.

First, to acknowledge the progress we had made, but more importantly to encourage further advances, we ended our blockage of Vietnam's access to international financial institutions.

Second, the President decided to send a new, high-level delegation to Vietnam to press for more progress on unresolved POW/MIA issues. I had the honor of co-leading that delegation, along with Deputy Secretary of Veterans Affairs Hershel Gober and Lt. Gen. Michael Ryan, assistant to the Chairman of the Joint Chiefs. We were accompanied by leading representatives of the four largest veterans organizations. Our mission was to ensure that Hanoi's top leaders understood the President's commitment to the POW/MIA issue. We stressed that further movement in bilateral relations required additional concrete results in four key areas identified by the President:

- Remains;
- Discrepancy cases;

- Trilateral cooperation with Laos and Vietnam; and
- Documents.

We also emphasized the importance we attach to human rights. We accomplished our mission. We delivered the President's message to the Party General Secretary, the Minister of Defense, the Acting Foreign Minister, and the Minister of the Interior. We also had a very productive session between veterans of both sides.

On September 13, the President decided to renew his authority to continue the embargo against Vietnam. However, to recognize POW/MIA progress in the four key areas and to stimulate further results, we modified the embargo to permit American companies to undertake development projects in Vietnam funded by international financial institutions.

In December, I returned to Vietnam to assess the overall situation, including progress in the four key areas. I held lengthy discussions with the outstanding personnel serving in our Joint Task Force. I met with Vietnam's Prime Minister, Foreign Minister, and other leaders. I traveled to the Vietnam-Laos border to observe first-hand the trilateral cooperation process there. I had the honor to witness the beginning of the journey home for the remains of missing Americans—a ceremony of stunning dignity that I will never forget.

In late December, the President's senior advisors met to review the POW/MIA record. They came to the unanimous conclusion that there had been significant, tangible progress in all four of the areas identified by the President in July.

The Criteria for Progress

What then were the results upon which the President's actions were based? Let me summarize the detailed information that was provided last week.

1. The recovery and repatriation of American remains: During the six months following the President's July announcement, we brought home the remains of 39 Americans—more than we repatriated in all of 1992. Throughout 1993, we repatriated the remains of 67

Americans, making last year the third-most-productive one for recovering remains since the end of the war. In the first month of this year, we have already brought home 12 more American remains.

2. The continued resolution of discrepancy cases and continued accomplishment of live sighting investigations: Since July 1993, we have confirmed the deaths of 19 individuals on our list of discrepancy cases. Since the beginning of the Administration, we have confirmed the deaths of 62 individuals, reducing the number of these cases from 135 to 73. We have a special team operating in Vietnam which is continuing to investigate the remaining 73. We have conducted more than 300 investigations on the ground in Vietnam of reported sightings of live American POWs and of cases of Americans who were last known to be alive during the war. None of these has produced evidence that an American POW is being held captive in Vietnam today. But we will continue to pursue vigorously any reports of live prisoners that we receive.

3. Further assistance in implementing trilateral investigations along the Vietnamese-Lao border: For many years we tried without success to investigate cases of Americans missing along the Vietnamese-Lao border, particularly airmen shot down over the Ho Chi Minh Trail. As a direct result of the President's July initiative, the Governments of Vietnam and Laos reached agreement in August 1993 to cooperate jointly on such investigations. The first such operation took place on the border of Vietnam and Laos in December. I personally visited with the Vietnamese, Lao, and American teams during my trip. The operation has succeeded in locating new remains as well as crash sites that we plan to excavate in the coming months.

4. Accelerated efforts to provide all POW/MIA-related documents that can give us answers to individual cases: Since July, we have received, for the first time, records from Vietnam's wartime anti-aircraft units along the Ho Chi Minh Trail. These records contain information about hundreds of U.S. airmen who were shot down and are listed as missing. This information should help us locate crash sites and recover remains in both Vietnam and Laos. We also have obtained, for the first time, documents from a wartime political-military unit. This material contains

information on American servicemen buried by North Vietnamese forces and written reports recounting unilateral efforts by Hanoi to locate the remains of Americans. This information should assist our efforts to achieve the fullest possible accounting.

Since the archival research program was initiated in October 1992, we have received from the Vietnamese 25,000 POW/MIA-related documents and artifacts. Of these, 600 have been correlated to unresolved cases. This represents more POW/MIA-related documentation than we had previously received during the entire period since the end of the Vietnam war. The President agreed with his advisers that this record represented "significant tangible progress."

Overall, we believe that 1993 was the most productive year for POW/MIA progress since the war.

Once again, I would like to pay special tribute to the incredible work being done by the men and women of the Joint Task Force–Full Accounting under the leadership of Adm. Charles Larson and Gen. Thomas Needham. They have endured hardships and dangers. They have displayed ingenuity, dedication, and tenacity in tracking down every possible lead. They deserve our utmost gratitude and respect. These men and women are a source of immense pride for all Americans. I also note the assistance we have received in the field from both official and private Vietnamese. Our Joint Task Force–Full Accounting personnel have reported that their cooperation during the past six months has been excellent. I cite two brief examples. In one instance, Vietnamese soldiers participating in an activity helped U.S. teams cross a Vietnam war–era minefield to an investigation site and helped remove four 100-pound bombs from a crater we wanted to excavate. In another incident, a U.S. excavation team had been working without success for two weeks when a local villager approached and said he had witnessed the wartime burial of an American. The villager then directed the team to the burial site, resulting in the recovery of remains for which the JTF-FA were searching. We now have in place the mechanisms we need to achieve the fullest possible accounting. We have the means to investigate any reports of possible live American prisoners. We have the mechanisms to excavate crash sites and burial locations. We have the means to interview witnesses in villages and Vietnam's wartime military leaders. We have special teams to search for remains and information on the highest priority discrepancy cases. We have mechanisms to review documents related to our missing men. And we have the means to investigate cases along the Vietnamese-Lao border. All of these instruments will help President Clinton fulfill his pledge to the families of the missing—that everything possible will be done to determine the fates of their missing fathers and sons, husbands and brothers. Let us also briefly recall two other positive aspects of recent U.S. engagement with Vietnam. As a result of the 1991 "roadmap" policy, Hanoi withdrew its troops from Cambodia and has supported the promising advance toward peace, freedom, and human rights in that country. The overwhelming turnout for last year's free Cambodian elections, in spite of intimidation and violence, clearly demonstrated that democracy is not only a Western ideal. Hanoi has also released from reeducation camps its citizens who had been detained because of their pre-1975 association with the United States or the former South Vietnamese Government. These developments are encouraging. So too are Vietnamese pledges of continued cooperation on POW/MIAs. But, as the President cautioned, "it must not end here." We will relentlessly continue our search for answers. We know from experience that this search will take a long time. Just two months ago we repatriated the remains of American aviators who were lost in World War II. Their remains, and the wreckage of their airplane, were found in the glaciers of Tibet, and returned with the cooperation and assistance of the Chinese Government. In recent months, we have also retrieved more remains and more answers concerning the Korean and Vietnam wars from North Korea, Russia, and China as well as the countries of Indochina. I am confident the Vietnamese understand the President's determination to see this issue through. They also know that any further steps in our relationship will depend on our making even more progress. Following the President's announcement, the Ministry of Foreign Affairs declared that the Vietnamese Government reiterates "its policy of consistently regarding the question of Americans missing from the war as a humanitarian concern not linked with political issues. . . . The Government and people of Vietnam have been, are, and will be cooperating in a constructive spirit with the American Government and people to solve this issue to the fullest possible extent."

When I informed Vietnam's Ambassador to the United Nations of the President's decisions, he said, "We promise to go forward with you to see the MIA issue resolved. I have a promise from Hanoi that cooperation will continue."

The Views of Others

The question for the President then was, what actions could we take to continue this important progress? How could we keep Vietnam motivated to pursue and expand its cooperation?

The President turned to many people for advice on these questions. He consulted with members of his cabinet most directly concerned with the POW/MIA issue, including the Secretary of State, the Secretary of Defense, and the National Security Adviser. The President asked General Shalikashvili and the Commander of our Pacific forces, Admiral Larson. He asked Gen. John Vessey, and the leaders of the delegation he sent to Hanoi last July. Everyone recommended that the best way to make more progress and resolve POW/MIA issues is to lift the embargo and expand our presence in Vietnam.

As I noted earlier, the President also sought the advice of many members of Congress. Here I would note the special contribution of Senator John Kerry, who co-chaired the Senate Select Committee on POW/MIA Affairs. For 15 months, the senator, a decorated veteran of Vietnam, steered his committee through an exhaustive investigation. The committee's findings played an important part in our deliberations. Senator Kerry also sponsored and championed the amendment endorsing an end to the embargo which the Senate overwhelmingly passed late last month. At the risk of not mentioning all of the amendment's co-sponsors, I do want to single out two others with particular backgrounds. In very personal and moving remarks on the floor, Senator John McCain, who spent almost six years as a POW in Vietnam, recommended ending the embargo as the best way to account finally for his missing brothers in arms. Senator Bob Kerrey, who

earned the Medal of Honor in Vietnam, also urged us to
end the embargo to resolve the POW/MIA issue, and to
make more progress on human rights and democracy
issues.

In the House of Representatives, I believe a broad
majority also supports the President's decisions. They
include many veterans and former POWs such as
Congressman Pete Peterson, who has served a central role
in our search for answers not only in Indochina but in the
former Soviet Union.

The President and other Administration officials also
consulted once again with the representatives of veterans
organizations and family groups. While many of them
disagree with the President's decision to lift the trade
embargo, they all share his objective of achieving the fullest
possible accounting. They agree in principle with the
strategic approach of the Administration—namely, to take
incremental steps forward in our relations with Vietnam in
response to progress and to encourage further progress.
They agree that Vietnamese activity has intensified in
recent months.

The disagreements arise over whether there has been
sufficient progress, as opposed to an extensive process, to
justify making another move forward. As I have outlined,
we believe that we have witnessed not only unprecedented
cooperation from the Vietnamese, but also substantial
tangible results from our joint efforts.

Despite these differences—and I don't wish to
minimize them—we look forward to working closely with
those who have the greatest personal stake in this difficult
issue. We welcome their continued counsel. We empathize
with their pain—not only over lost family members and
comrades, but over the past deceptions by the Vietnamese
and inadequate performance by the U.S. Government.

After considering all views, the President made his
decisions. In short, he agrees with all his senior advisers,
with our military personnel working on the ground, and
with an overwhelming bipartisan majority in the Congress
that the actions he announced represent the best way to
account for our missing men. The steps we have taken do
not represent full "normalization" of relations with
Vietnam. We are not opening embassies or exchanging
ambassadors. We are not granting Vietnam special

economic privileges. We retain considerable political and economic incentives to ensure that the government of Vietnam does not waver from its commitment to continue its cooperation on POW/MIA issues. Our efforts will continue undiminished, indeed with fresh momentum.

With these prospects in mind, President Clinton also decided to establish a liaison office in Vietnam and to permit the Vietnamese to open a similar office here. We believe such offices will greatly assist in our search for MIA information. They will also serve to expand our dialogue with Vietnam on many issues, including human rights. And they will support and protect American visitors, tourists, and business people.

The vastly increased numbers of American visitors, tourists, business people, and other private groups who will now spread across Vietnam should produce greater openness, greater contacts, greater information on our MIAs—and concrete results.

At this moment, we are only in the initial planning stages for the liaison offices. Questions on timing, staffing, privileges and immunities, and functions will be the subject of discussions with the Vietnamese. We plan to begin these talks in the near future. We welcome your views and will keep you and your staff apprised of significant developments.

In sum, President Clinton and all of his top advisers believe that it is time to acknowledge the help of the Vietnamese. The February 3, 1994, decisions will encourage further efforts by demonstrating to Vietnamese leaders and the Vietnamese people that we will meet cooperation with reciprocal steps, that it is in their interest to continue helping us. The families and loved ones of our missing Americans deserve answers. The President's actions mark a major milestone on a lengthy journey in pursuit of that goal. They represent a new beginning, a rededication to our ongoing labors.

Human Rights

Before concluding, let me cite two other important issues with Vietnam. My colleagues and I have raised these subjects regularly, including at the highest levels in Hanoi, and in Secretary [of State William] Christopher's meeting last fall with Deputy Prime Minister Khai. The first issue concerns American citizens who are incarcerated in Vietnam. We know of five such Americans, and are disappointed that we have only been granted access to one of them. Now, with the opening of an official U.S. office in Hanoi, we expect our discussions with Vietnam to lead to normal consular access in accordance with international practice and law. The second issue is human rights. The just-published State Department 1993 human rights report for Vietnam spells out our deep concerns. It states, in part, that the Vietnamese Government "continued to violate human rights in 1993. The authorities continued to limit severely freedom of speech, press, assembly and association, as well as worker rights and the fight of citizens to change their government."

In my December meetings in Hanoi, Vietnam agreed to hold regular bilateral discussions with us on human rights. These should commence later this month. We expect a constructive, productive forum in which we will continue to urge Hanoi to respect universal human rights, and release those detained for the peaceful expression of political or religious beliefs. I would note here our sustained, personal concern for the health of Dr. Nguyen Dan Que, among others. We have raised his case on many occasions, most recently in my meeting last week with the Vietnamese ambassador to the United Nations. We will continue to follow closely his fate and that of others in similar situations. The further exposure of Vietnamese society to outside trade, investment, people, information, and ideas as a result of the President's decisions, should work to open up the political system of Vietnam.

Vietnam clearly has far to go to improve its observance of human rights. Some actions by the Vietnamese leadership in recent years, however, have signaled their intention to reintegrate their nation in the world and contribute to the stability of the Southeast Asian region.

As I already noted, the Vietnamese were a signatory to the Cambodia Peace Accords and have faithfully supported the implementation of the peace process. While the government's institution of economic reforms is clearly in Vietnam's self-interest, it has also had a positive impact on the region and drawn that nation more into the world trading community. The Vietnamese have also demonstrated a willingness to resolve their territorial disputes in the South

China Sea with China and other Southeast Asian claimants in a peaceful and constructive manner.

Conclusion

Let me close with the words of President Clinton last Thursday:

"Whatever the Vietnam war may have done in dividing our country in the past, today our nation is one in honoring those who served and pressing for answers about all those who did not return. This decision today, I believe, renews that commitment and our constant, constant effort never to forget those until our job is done. Those who have sacrificed deserve a full and final accounting. I am absolutely convinced, as are so many in Congress who served there and so many Americans who have studied the issue, that this decision today will help to ensure that fullest possible accounting."

Mr. Chairman, members of this committee: As we look back upon this time many years from now, perhaps the most significant dimension of the President's decision will prove to be psychological. Perhaps we have begun turning the pages of history. Perhaps we are moving toward eventual reconciliation with a former enemy. Perhaps for Americans, as one observer has put it, Vietnam will become a country, not a war. Perhaps we are further developing the President's vision of a New Pacific Community. Above all, let us hope that—whatever our differences about the war or how to re-solve its lingering questions—we have truly advanced the process of healing the wounds. May the families at last find answers. And may all Americans at last find peace.

Source: Lord 1994, 105–111.

Senator Bob Dole's Speech against Normalization, 10 July 1995

The day before President Clinton announced the normalization of rela-tions with Vietnam, Senate Majority Leader Bob Dole expressed the reservations of many on this issue before the Senate.

News reports indicate that President Clinton is on the verge of making a decision about normalizing relations with Vietnam. I understand an announcement may come as soon as tomorrow. Secretary of State Warren Christopher has recommended normalization. Many Vietnam veterans support normalization—including a bipartisan group of veterans in the Senate, led by the senior Senator from Arizona, John McCain. Many oppose normalization as well. Just as the Vietnam war divided Americans in the 1960s and 1970s, the issue of how to finalize peace with Vietnam divides Americans today.

At the outset, let me observe that there are men and women of good will on both sides of this issue. No one should question the motives of advocates or opponents of normalization. We share similar goals: Obtaining the fullest possible accounting for American prisoners of war and missing in action; continuing the healing process in the aftermath of our most divisive war; fostering respect for human rights and political liberty in Vietnam. . . .

The debate over normalization is about our differences with the Government of Vietnam, not with the Vietnamese people. The people of Vietnam have suffered decades of war and brutal dictatorship. We hope for a better future for the people of Vietnam—a future of democracy and freedom, not repression and despair.

The debate over normalization is not a debate over the ends of American policy; it is a debate over the means. The most fundamental question is whether normalizing relations with Vietnam will further the goals we share. In my view, now is not the time to normalize relations with Vietnam. The historical record shows that Vietnam cooperates on POW/MIA issues only when pressured by the United States; in the absence of sustained pressure, there is little progress on POW/MIA concerns, or on any other issue.

The facts are clear. Vietnam is still a one party Marxist dictatorship. Preserving their rule is the No. 1 priority of Vietnam's Communist Government. Many credible sources suggest Vietnam is not providing all the information it can on POW/MIA issues. In some cases, increased access has only confirmed how much more Vietnam could be doing. This is not simply my view, it is a view shared by two Asia experts—Steve Solarz, former chairman of the House Subcommittee on Asia and Pacific Affairs, and Richard Childress, National Security Council Vietnam expert from

1981 to 1989. Earlier this year, they wrote: "Vietnam could easily account for hundreds of Americans by a combination of unilateral repatriation of remains, opening its archives, and full cooperation on U.S. servicemen missing in Laos."

Again, not my quote but a quote by the two gentlemen mentioned. They conclude that "whatever the reasons or combination of reasons, Vietnam, in the current environment, has made a conscious decision to keep the POW/MIA issue alive by not resolving it."

This is a view shared by the National League of POW/MIA Families, which has worked tirelessly to resolve this issue for many years. It is also a view shared by major veterans groups, including the American Legion, the largest veterans group. The media have reported that the Veterans of Foreign Wars, the second largest group, is supportive of normalization. Let me quote from VFW's official position adopted at its 1994 convention: "At some point in time but only after significant results have been achieved through Vietnam/U.S. cooperative efforts, we should . . . move towards normalizing diplomatic relations."

A more recent VFW statement makes clear that normalization is not opposed by the VFW if it leads to a fuller accounting of POW/MIA cases.

If President Clinton intends to normalize diplomatic relations with Vietnam, he should do so only after he can clearly state that Vietnam has done everything it reasonably can to provide the fullest possible accounting. That is the central issue. The United States has diplomatic relations with many countries which violate human rights, and repress their own people. But the United States should not establish relations with a country which withholds information about the fate of American servicemen. . . .

No doubt about it, the Vietnamese Government wants normalization very badly. Normalization is the strongest bargaining chip America has. As such, it should only be granted when we are convinced Vietnam has done all it can do. Vietnam has taken many steps—sites are being excavated, and some remains have been returned. But there are also signs that Vietnam may be willfully withholding information. Unless the President is absolutely convinced Vietnam has done all it can do to resolve the POW/MIA issue—and is willing to say so publicly and unequivocally —it would be a strategic, diplomatic and moral mistake to grant Vietnam the stamp of approval from the United States.

Source: Dole 1995, S9626–S9628.

President Clinton's Speech on Normalizing Relations, 11 July 1995

President Clinton made the following statement as he announced the normalization of relations with the Socialist Republic of Vietnam. In his speech, he cited the progress made, called normalization the "next appropriate step," and pledged to continue accounting efforts.

Thank you very much. I welcome you all here. . . .

Today, I am announcing the normalization of diplomatic relations with Vietnam. From the beginning of this Administration, any improvement in the relationship between America and Vietnam has depended upon making progress on the issue of Americans who were missing in action or held as prisoners of war. Last year, I lifted the trade embargo on Vietnam in response to their cooperation and to enhance our efforts to secure the remains of lost Americans and to determine the fate of those whose remains have not been found.

It has worked. In 17 months, Hanoi has taken important steps to help us resolve many cases. Twenty-nine families have received the remains of their loved ones and at last have been able to give them a proper burial. Hanoi has delivered to us hundreds of pages of documents shedding light on what happened to Americans in Vietnam, and Hanoi has stepped up its cooperation with Laos, where many Americans were lost.

We have reduced the number of so-called discrepancy cases—in which we have had reason to believe that Americans were still alive after they were lost—to 55. And we will continue to work to resolve more cases.

Hundreds of dedicated men and women are working on all these cases, often under extreme hardship and real dangers in the mountains and jungles of Indochina. On behalf of all Americans, I want to thank them. And I want to pay a special tribute to Gen. John Vessey, who has worked so tirelessly on this issue for Presidents Reagan and Bush and for our Administration. He has made a great difference to a great many families and we as a nation are grateful for his dedication and for his service. Thank you, sir. I also want to thank the presidential delegation, led by Deputy Secretary of Veterans Affairs Hershel Gober, Winston Lord, and James Wold, who have helped us to make so much progress on this issue. I am especially grateful to the leaders of the families and the veterans organizations who have worked with the delegation and maintained their extraordinary commitment to finding the answers we seek.

Never before in the history of warfare has such an extensive effort been made to resolve the fate of soldiers who did not return. Let me emphasize, normalization of our relations with Vietnam is not the end of our effort. From the early days of this Administration, I have said to the families and veterans groups what I say again here: We will keep working until we get all the answers we can. Our strategy is working. Normalization of relations is the next appropriate step. With this new relationship, we will be able to make more progress. To that end, I will send another delegation to Vietnam this year, and Vietnam has pledged it will continue to help us find answers. We will hold them to that pledge.

By helping to bring Vietnam into the community of nations, normalization also serves our interest in working for a free and peaceful Vietnam in a stable and peaceful Asia. We will begin to normalize our trade relations with Vietnam, whose economy is now liberalizing and integrating into the economy of the Asia-Pacific region. Our policy will be to implement the appropriate United States Government programs to develop trade with Vietnam consistent with U.S. law.

As you know, many of these programs require certifications regarding human rights and labor rights before they can proceed. We have already begun discussing human rights issues with Vietnam, especially issues regarding religious freedom. Now we can expand and strengthen that dialogue. The Secretary of State will go to Vietnam in August where he will discuss all of these issues, beginning with our POW and MIA concerns.

I believe normalization and increased contact between Americans and Vietnamese will advance the cause of freedom in Vietnam, just as it did in Eastern Europe and the former Soviet Union. I strongly believe that engaging the Vietnamese on the broad economic front of economic reform and the broad front of democratic reform will help to honor the sacrifice of those who fought for freedom's sake in Vietnam.

I am proud to be joined in this view by distinguished veterans of the Vietnam War. They served their country bravely. They are of different parties. A generation ago they had different judgments about the war which divided us so deeply. But today, they are of a single mind. They agree that the time has come for America to move forward on Vietnam. All Americans should be grateful especially that Senators John McCain, John Kerry, Bob Kerrey, and Chuck Robb; Representative Pete Peterson, along with other Vietnam veterans in the Congress, including Senator Harkin, Congressman Colby, and Congressman Gilchrist, who just left; and others who are out here in the audience have kept up their passionate interest in Vietnam, but were able to move beyond the haunting and painful past toward finding common ground for the future. Today, they and many other veterans support normalization of relations, giving the opportunity to Vietnam to fully join the community of nations and being true to what they fought for so many years ago.

Source: Clinton 1995b, 551.

Proclamation 6818: National POW/MIA Recognition Day, 1995

The presidential proclamation below, dated 29 August 1995, is one of many that has over the years acknowledged the sacrifices made by prisoners of war and MIAs and their families.

Throughout our proud history, America's sons and daughters have answered the call to defend our fundamental liberties and to safeguard the freedoms of peace-seeking countries around the globe. Representing the finest this Nation has to offer, the members of our Armed Forces have given everything of themselves in defense of the independence and democracy that we hold so dear.

This year we have a special opportunity to honor their service as we commemorate the 50th anniversary of the end of World War II, the dedication of the Korean War Veterans Memorial, and the unveiling of the POW and MIA postage stamp.

In remembering these heroic men and women, it is with profound respect and solemn appreciation that we single out those who paid the heaviest price. Among them are the Prisoners of War and those Missing in Action. Their courage and devotion to duty, honor, and country—often in the face of brutal treatment and torture by their captors— will never be forgotten by the American people.

Our Nation also recognizes that the families of these brave citizens have suffered and made great sacrifices for our country. For it is in the name of both the missing and their loved ones that we aggressively pursue the release of any United States service member held against his or her will, that we search tirelessly for information about the missing, and that we seek the repatriation of recoverable American remains.

On September 15, 1995, the flag of the National League of Families of American Prisoners of War and Missing in Southeast Asia, a black and white banner symbolizing America's missing, will be flown over the White House, the Capitol, the United States Departments of State, Defense, and Veterans Affairs, the Selective Service System Headquarters, the Vietnam Veterans and Korean War Veterans Memorials, and national cemeteries across the country. This flag is a symbol of our Nation's covenant with those who defend us and with the loved ones they leave behind—the brave individuals who have earned our everlasting gratitude and their families who deserve our deepest sympathy and our national pledge to achieve the fullest possible accounting of American troops.

Now, Therefore, I, William J. Clinton, President of the United States of America, by virtue of the authority vested in me by the Constitution and laws of the United States, do hereby proclaim September 15, 1995, as "National POW/MIA Recognition Day." I urge State and local officials, private organizations, and citizens everywhere to join in honoring all Prisoners of War and Missing in Action still unaccounted for as a result of their dedicated service to our great country. I also encourage the American people to recognize and acknowledge the steadfast vigil the families of the missing maintain in their quest for answers and a conclusion to their struggle. Finally, I call upon all Americans to observe this day with appropriate ceremonies and activities. In Witness Whereof, I have hereunto set my hand this twenty-ninth day of August, in the year of our Lord nineteen hundred and ninety-five, and of the Independence of the United States of America the two hundred and twentieth.

Source: Clinton 1995a, 1465.

Treaties and Legislation

The Geneva Convention for the Protection of War Victims: Armed Forces in the Field (1949)

The Geneva Conventions of 1949, of which a portion is excerpted below, is made up of four conventions: Convention for the Amelioration of the Condition of the Wounded and Sick in Armed Forces in the Field; Convention for the Amelioration of the Condition of the Wounded, Sick and Shipwrecked Members of Armed Forces at Sea; Convention Relative to the Treatment of Prisoners of War; and Convention for the Protection of Civilian Persons in Time of War. Like the 1929 convention that preceded it, these conventions were meant to make warfare between nations a more "humane" process. After the 1929 Geneva Convention was disregarded during World War II by the Germans and Japanese, representatives from many nations reconvened to reaffirm the previous conventions. Among the behaviors forbidden by the 1949 Geneva Convention were the taking of hostages, torture, and collective punishment or reprisals. Prisoners were to be treated humanely, which included being given adequate food, medical treatment, relief supplies, and mail. The North Vietnamese, maintaining that American POWs were war criminals involved in an illegal war (no formal declaration of war was ever made by the United States), chose not to abide by the Geneva Conventions.

[Convention, with annexes, dated August 12, 1949, at Geneva, Switzerland; ratification advised by the Senate of the United States of America, subject to a reservation and statement, July 6, 1955; ratified by the president of the United States of America, subject to the said reservation and statement, July 14, 1955; ratification of the United States of America deposited with the Swiss Federal Council on August 2, 1955; proclaimed by the president of the United States of America on August 30, 1955; date of entry into force with respect to the United States of America: February 2, 1956.]

By the President of the United States of America:

A Proclamation

WHEREAS the Geneva Convention for the Amelioration of the Condition of the Wounded and Sick in Armed Forces in the Field was open for signature from August 12, 1949, until February 12, 1950, and during that period was signed on behalf of the United States of America and sixty other States; . . .

The undersigned Plenipotentiaries of the Governments represented at the Diplomatic Conference held at Geneva from April 21 to August 12, 1949, for the purpose of revising the Geneva Convention for the Relief of the Wounded and Sick in Armies in the Field of July 27, 1929, have agreed as follows:

CHAPTER I

GENERAL PROVISIONS

Article 1

The High Contracting Parties undertake to respect and to ensure respect for the present Convention in all circumstances.

Article 2

In addition to the provisions which shall be implemented in peacetime, the present Convention shall apply to all cases of declared war or of any other armed conflict which may arise between two or more of the High Contracting Parties, even if the state of war is not recognized by one of them.

The Convention shall also apply to all cases of partial or total occupation of the territory of a High Contracting Party, even if the said occupation meets with no armed resistance. Although one of the Powers in conflict may not be a party to the present Convention, the Powers who are parties thereto shall remain bound by it in their mutual relations. They shall furthermore be bound by the Convention in relation to the said Power, if the latter accepts and applies the provisions thereof.

Article 3

In the case of armed conflict not of an international character occurring in the territory of one of the High Contracting Parties, each Party to the conflict shall be bound to apply, as a minimum, the following provisions:

(1) Persons taking no active part in the hostilities, including members of armed forces who have laid down their arms and those placed hors de combat by sickness, wounds, detention, or any other cause, shall in all circumstances be treated humanely, without any adverse distinction founded on race, colour, religion or faith, sex, birth or wealth, or any other similar criteria.

To this end, the following acts are and shall remain prohibited at any time and in any place whatsoever with respect to the above-mentioned persons:

(a) violence to life and person, in particular murder of all kinds, mutilation, cruel treatment and torture;

(b) taking of hostages;

(c) outrages upon personal dignity, in particular humiliating and degrading treatment;

(d) the passing of sentences and the carrying out of executions without previous judgment pronounced by a regularly constituted court, affording all the judicial guarantees which are recognized as indispensable by civilized peoples.

(2) The wounded and sick shall be collected and cared for. An impartial humanitarian body, such as the International Committee of the Red Cross, may offer its services to the Parties to the conflict.

The Parties to the conflict should further endeavour to bring into force, by means of special agreements, all or part of the other provisions of the present Convention. The application of the preceding provisions shall not affect the legal status of the Parties to the conflict.

Article 4

Neutral Powers shall apply by analogy the provisions of

the present Convention to the wounded and sick, and to members of the medical personnel and to chaplains of the armed forces of the Parties to the conflict, received or interned in their territory, as well as to dead persons found.

Article 5

For the protected persons who have fallen into the hands of the enemy, the present Convention shall apply until their final repatriation.

Article 6

In addition to the agreements expressly provided for in Articles 10, 15, 23, 28, 31, 36, 37 and 52, the High Contracting Parties my conclude other special agreements for all matters concerning which they may deem it suitable to make separate provision. No special agreement shall adversely affect the situation of the wounded and sick, of members of the medical personnel or of chaplains, as defined by the present Convention, nor restrict the rights which it confers upon them.

Wounded and sick, as well as medical personnel and chaplains, shall continue to have the benefit of such agreements as long as the Convention is applicable to them, except where express provisions to the contrary are contained in the aforesaid or in subsequent agreements, or where more favourable measures have been taken with regard to them by one or other of the Parties to the conflict.

Article 7

Wounded and sick, as well as members of the medical personnel and chaplains, may in no circumstances renounce in part or in entirety the rights secured to them by the present Convention, and by the special agreements referred to in the foregoing Article, if such there be.

Article 8

The present Convention shall be applied with the cooperation and under the scrutiny of the Protecting Powers whose duty it is to safeguard the interests of the

Parties to the conflict. For this purpose, the Protecting Powers may appoint, apart from their diplomatic or consular staff, delegates from amongst their own nationals or the nationals of other neutral Powers. The said delegates shall be subject to the approval of the Power with which they are to carry out their duties.

The Parties to the conflict shall facilitate to the greatest extent possible, the task of the representatives or delegates of the Protecting Powers. The representatives or delegates of the Protecting Powers shall not in any case exceed their mission under the present Convention. They shall, in particular, take account of the imperative necessities of security of the State wherein they carry out their duties. Their activities shall only be restricted as an exceptional and temporary measure when this is rendered necessary by imperative military necessities.

Article 9

The provisions of the present Convention constitute no obstacle to the humanitarian activities which the International Committee of the Red Cross or any other impartial humanitarian organization may, subject to the consent of the Parties to the conflict concerned, undertake for the protection of wounded and sick, medical personnel and chaplains, and for their relief.

Article 10

The High Contracting Parties may at any time agree to entrust to an organization which offers all guarantees of impartiality and efficacy the duties incumbent on the Protecting Powers by virtue of the present Convention.

When wounded and sick, or medical personnel and chaplains do not benefit or cease to benefit, no matter for what reason, by the activities of a Protecting Power or of an organization provided for in the first paragraph above, the Detaining Power shall request a neutral State, or such an organization, to undertake the functions performed under the present Convention by a Protecting Power designated by the Parties to a conflict.

If protection cannot be arranged accordingly, the Detaining Power shall request or shall accept, subject to the

provisions of this Article, the offer of the services of a
humanitarian organization, such as the International
Committee of the Red Cross, to assume the humanitarian
functions performed by Protecting Powers under the
present Convention.

Any neutral Power, or any organization invited by the
Power concerned or offering itself for these purposes, shall
be required to act with a sense of responsibility towards
the Party to the conflict on which persons protected by the
present Convention depend, and shall be required to
furnish sufficient assurances that it is in a position to
undertake the appropriate functions and to discharge them
impartially.

No derogation from the preceding provisions shall be
made by special agreements between Powers one of which
is restricted, even temporarily, in its freedom to negotiate
with the other Power or its allies by reason of military
events, more particularly where the whole, or a substantial
part, of the territory of the said Power is occupied.

Whenever, in the present Convention, mention is made
of a Protecting Power, such mention also applies to substi-
tute organizations in the sense of the present Article.

Article 11

In cases where they deem it advisable in the interest of protected persons, particularly in cases of disagreement
between the Parties to the conflict as to the application or
interpretation of the provisions of the present Convention,
the Protecting Powers shall lend their good offices with a
view to settling the disagreement. For this purpose, each of
the Protecting Powers may, either at the invitation of one
Party or on its own initiative, propose to the Parties to the
conflict a meeting of their representatives, in particular of
the authorities responsible for the wounded and sick,
members of medical personnel and chaplains, possibly on
neutral territory suitably chosen. The Parties to the conflict
shall be bound to give effect to the proposals made to them
for this purpose. The Protecting Powers may, if necessary,
propose for approval by the Parties to the conflict, a person
belonging to a neutral Power or delegated by the
International Committee of the Red Cross, who shall be
invited to take part in such a meeting.

CHAPTER II

WOUNDED AND SICK

Article 12

Members of the armed forces and other persons mentioned in the following Article, who are wounded or sick, shall be respected and protected in all circumstances.

They shall be treated humanely and cared for by the Party to the conflict in whose power they may be, without any adverse distinction founded on sex, race, nationality, religion, political opinions, or any other similar criteria. Any attempts upon their lives or violence to their persons shall be strictly prohibited; in particular, they shall not be murdered or exterminated, subjected to torture or to biological experiments; they shall not wilfully be left without medical assistance and care, nor shall conditions exposing them to contagion or infection be created.

Only urgent medical reasons will authorize priority in the order of treatment to be administered.

Women shall be treated with consideration due to their sex. The Party to the conflict which is compelled to abandon wounded or sick to the enemy shall, as far as military considerations permit, leave with them a part of its medical personnel and material to assist in their care.

Article 13

The present Convention shall apply to the wounded and sick belonging to the following categories:

(1) Members of the armed forces of a Party to the conflict, as well as members of militias or volunteer corps forming part of such armed forces.

(2) Members of other militias and members of other volunteer corps, including those of organized resistance movements, belonging to a Party to the conflict and operating in or outside their own territory, even if this territory is occupied, provided that such militias or volunteer corps, including such organized resistance movements, fulfil the following conditions:

(a) that of being commanded by a person responsible for his subordinates;

(b) that of having a fixed distinctive sign recognizable at a distance;

(c) that of carrying arms openly;

(d) that of conducting their operations in accordance with the laws and customs of war.

(3) Members of regular armed forces who profess allegiance to a Government or an authority not recognized by the Detaining Power.

(4) Persons who accompany the armed forces without actually being members thereof, such as civil members of military aircraft crews, war correspondents, supply contractors, members of labour units or of services responsible for the welfare of the armed forces, provided that they have received authorization from the armed forces which they accompany.

(5) Members of crews, including masters, pilots and apprentices, of the merchant marine and the crews of civil aircraft of the Parties to the conflict, who do not benefit by more favourable treatment under any other provisions in international law.

(6) Inhabitants of a non-occupied territory who, on the approach of the enemy, spontaneously take up arms to resist the invading forces, without having had time to form themselves into regular armed units, provided they carry arms openly and respect the laws and customs of war.

Article 14

Subject to the provisions of Article 12, the wounded and sick of a belligerent who fall into enemy hands shall be prisoners of war, and the provisions of international law concerning prisoners of war shall apply to them.

Article 15

At all times, and particularly after an engagement, Parties to the conflict shall, without delay, take all possible measures to search for and collect the wounded and sick, to protect them against pillage and ill-treatment, to ensure their adequate care, and to search for the dead and prevent their being despoiled. Whenever circumstances permit, an armistice or a suspension of fire shall be arranged, or local arrangements made, to permit the removal, exchange and transport of the wounded left on the battlefield. Likewise, local arrangements may be concluded between Parties to the conflict for the removal or exchange of wounded and sick from a besieged or encircled area, and for the passage of medical and religious personnel and equipment on their way to that area.

Article 16

Parties to the conflict shall record as soon as possible, in respect of each wounded, sick or dead person of the adverse Party falling into their hands, any particulars which may assist in his identification. These records should if possible include:

(a) designation of the Power on which he depends;

(b) army, regimental, personal or serial number;

(c) surname;

(d) first name or names;

(e) date of birth;

(f) any other particulars shown on his identity card or disc;

(g) date and place of capture or death;

(h) particulars concerning wounds or illness, or cause of death.

As soon as possible the above-mentioned information shall be forwarded to the Information Bureau described in Article 122 of the Geneva Convention relative to the Treatment of Prisoners of War of August 12, 1949, which shall transmit this information to the Power on which these persons depend through the intermediary of the Protecting Power and of the Central Prisoners of War Agency.

Parties to the conflict shall prepare and forward to each other through the same bureau, certificates of death or duly authenticated lists of the dead. They shall likewise collect and forward through the same bureau one half of a double identity disc, last wills or other documents of importance to the next of kin, money and in general all articles of an intrinsic or sentimental value, which are found on the dead. These articles, together with unidentified articles, shall be sent in sealed packets, accompanied by statements giving all particulars necessary for the identification of the deceased owners, as well as by a complete list of the contents of the parcel.

Article 17

Parties to the conflict shall ensure that burial or cremation of the dead, carried out individually as far as circumstances permit, is preceded by a careful examination, if possible by a medical examination, of the bodies, with a view to confirming death, establishing identity and enabling a report to be made. One half of the double identity disc, or the identity disc itself if it is a single disc, should remain on the body. Bodies shall not be cremated except for imperative reasons of hygiene or for motives based on the religion of the deceased. In case of cremation, the circumstances and reasons for cremation shall be stated in detail in the death certificate or on the authenticated list of the dead.

They shall further ensure that the dead are honourably interred, if possible according to the rites of the religion to which they belonged, that their graves are respected, grouped if possible according to the nationality of the deceased, properly maintained and marked so that they may always be found. For this purpose, they shall organize at the commencement of hostilities an Official Graves Registration Service, to allow subsequent exhumations and to ensure the identification of bodies, whatever the site of the graves, and the possible transportation to the home country. These provisions shall likewise apply to the ashes, which shall be kept by the Graves Registration Service until proper disposal thereof in accordance with the wishes of the home country. As soon as circumstances permit, and at latest at the end of hostilities, these Services shall exchange, through the Information Bureau mentioned in the second paragraph of Article 16, lists showing the exact location and markings of the graves, together with particulars of the dead interred therein.

Article 18

The military authorities may appeal to the charity of the inhabitants voluntarily to collect and care for, under their direction, the wounded and sick, granting persons who have responded to this appeal the necessary protection and facilities. Should the adverse Party take or retake control of the area, he shall likewise grant these persons the same protection and the same facilities.

The military authorities shall permit the inhabitants and relief societies, even in invaded or occupied areas, spontaneously to collect and care for wounded or sick of whatever nationality. The civilian population shall respect these wounded and sick, and in particular abstain from offering them violence.

No one may ever be molested or convicted for having nursed the wounded or sick. The provisions of the present Article do not relieve the occupying Power of its obligation to give both physical and moral care to the wounded or sick.

Source: United States Treaties and Other International Agreements 1956, 3114–3128.

Paris Peace Accords (1973)

"The Agreement on Ending the War and Restoring Peace in Viet-Nam" was commonly known as the Paris Peace Accords. On 17 January 1973, this agreement ended U.S. participation in the war in Vietnam and set the procedures for the withdrawal of American military forces and the repatriation of American prisoners of war.

FOR THE GOVERNMENT OF THE UNITED STATES OF AMERICA:
  William P. Rogers
  Secretary of State

FOR THE GOVERNMENT OF THE REPUBLIC OF VIET-NAM:
  Tran Van Lam
  Minister for Foreign Affairs

FOR THE GOVERNMENT OF THE DEMOCRATIC REPUBLIC OF VIETNAM:
  Nguyen Duy Trinh
  Minister for Foreign Affairs

FOR THE PROVISIONAL REVOLUTIONARY GOVERNMENT OF THE REPUBLIC OF SOUTH VIET-NAM:
  Nguyen Thi Binh
  Minister of Foreign Affairs

Protocol to the Agreement on Ending the War and Restoring Peace in Viet-Nam Concerning the Return of Captured Military Personnel and Foreign Civilians and Captured and Detained Vietnamese Civilian Personnel

The Parties participating in the Paris Conference on Viet-Nam

In implementation of Article 8 of the Agreement on Ending the War and Restoring Peace in Viet-Nam signed on this date providing for the return of captured military personnel and foreign civilians, and captured and detained Vietnamese civilian personnel,

Have agreed as follows:

The Return of Captured Military Personnel and Foreign Civilians

Article 1

The parties signatory to the Agreement shall return the captured military personnel of the parties mentioned in Article 8

(a) of the agreement as follows:

—all captured military personnel of the United States and those of the other foreign countries mentioned in Article 3 (a) of the Agreement shall be returned to the United States authorities;

—all captured Vietnamese military personnel, whether belonging to regular or irregular armed forces, shall be returned to the two South Vietnamese parties; they shall be returned to that South Vietnamese party under whose command they served.

Article 2

All captured civilians who are nationals of the United States or of any other foreign countries mentioned in Article 3 (a) of the Agreement shall be returned to the United States authorities. All other captured foreign civilians shall be returned to the authorities of their country of nationality by any one of the parties willing and able to do so.

Article 3

The parties shall today exchange complete lists of captured persons mentioned in Articles 1 and 2 of this Protocol.

Article 4

(a) The return of all captured persons mentioned in Articles 1 and 2 of this Protocol shall be completed within sixty days of the signing of the Agreement at a rate no slower than the rate of withdrawal from South Viet-Nam of United States forces and those of the other foreign countries mentioned in Article 5 of the Agreement.

Article 5

The return and reception of the persons mentioned in Articles 1 and 2 of this Protocol shall be carried out at places convenient to the concerned parties. Places of return shall be agreed upon by the Four-Party Joint Military Commission. The parties shall ensure the safety of personnel engaged in the return and reception of those persons.

Article 6

Each party shall return all captured persons mentioned in Articles 1 and 2 of this Protocol without delay and shall facilitate their return and reception. The detaining parties shall not deny or delay their return for any reason, including the fact that captured persons may, on any grounds, have been prosecuted or sentenced.

The Return of Captured and Detained Vietnamese Civilian Personnel

Article 7

(a) The question of the return of Vietnamese civilian personnel captured and detained in South Viet-Nam will be resolved by the two South Vietnamese parties on the basis of the principles of Article 21 (b) of the Agreement on the Cessation of Hostilities in Viet-Nam of July 20, 1954, which reads as follows:

"The term 'civilian internees' is understood to mean all persons who, having in any way contributed to the political and armed struggle between the two parties, have been arrested for that reason and have been kept in detention by either party during the period of hostilities."

(b) The two South Vietnamese parties will do their utmost to resolve this question within ninety days after the cease-fire comes into effect.

(c) Within fifteen days after the cease-fire comes into effect, the two South Vietnamese parties shall exchange lists of the Vietnamese civilian personnel captured and detained by each party and lists of the places at which they are held.

Treatment of Captured Persons during Detention

Article 8

(a) All captured military personnel of the parties and captured foreign civilians of the parties shall be treated humanely at all times, and in accordance with international practice. They shall be protected against all violence to life and person, in particular against murder in any form, mutilation, torture and cruel treatment, and outrages upon personal dignity. These persons shall not be forced to join the armed forces of the detaining party.

They shall be given adequate food, clothing, shelter, and the medical attention required for their state of health. They shall be allowed to exchange post cards and letters with their families and receive parcels.

(b) All Vietnamese civilian personnel captured and detained in South Viet-Nam shall be treated humanely at all times, and in accordance with international practice. They shall be protected against all violence to life and person, in particular against murder in any form, mutilation, torture and cruel treatment, and outrages upon personal dignity. The detaining parties shall not deny or delay their return for any reason, including the fact that captured persons may, on any grounds, have been prosecuted or sentenced. These persons shall not be forced to join the armed forces of the detaining party.

They shall be given adequate food, clothing, shelter, and the medical attention required for their state of health. They shall be allowed to exchange post cards and letters with their families and receive parcels.

Article 9

(a) To contributed to improving the living conditions of the captured military personnel of the parties and foreign civilians of the parties, the parties shall, within fifteen days after the cease-fire comes into effect, agree upon the designation of two or more national Red Cross societies to visit all places where captured military personnel and foreign civilians are held.

(b) To contribute to improving the living conditions of

the captured and detained Vietnamese civilian personnel, the two South Vietnamese parties shall, within fifteen days after the cease-fire comes into effect, agree upon the designation of two or more national Red Cross societies to visit all places where the captured and detained Vietnamese civilian personnel are held.

With Regard to Dead and Missing Persons

Article 10

(a) The Four-Party Joint Military Commission shall ensure joint action by the parties in implementing Article 8 (b) of the Agreement. When the Four-Party Joint Military Commission has ended its activities, a Four-Party Joint Military team shall be maintained to carry on this task.

(b) With regard to Vietnamese civilian personnel dead or missing in South Viet-Nam, the two South Vietnamese parties shall help each other to obtain information about missing persons, determine the location and take care of the graves of the dead, in a spirit of national reconciliation and concord, in keeping with the people's aspirations.

Other Provisions

Article 11

(a) The Four-Party and Two-Party Joint Military Commissions will have the responsibility of determining immediately the modalities of implementing the provisions of this Protocol consistent with their respective responsibilities under Articles 16 (a) and 17 (a) of the Agreement. In case the Joint Military Commissions, when carrying out their tasks, cannot reach agreement on a matter pertaining to the return of captured personnel they shall refer to the International Commission for its assistance.

(b) The Four-Party Joint Military Commission shall form, in addition to the teams established by the Protocol concerning the cease-fire in South Viet-Nam and the Joint Military Commissions, a subcommission on captured persons and, as required, joint military teams on captured persons to assist the Commission in its tasks.

(c) From the time the cease-fire comes into force to the

time when the Two-Party Joint Military Commission becomes operational, the two South Vietnamese parties' delegations to the Four-Party Joint Military Commission shall form a provisional sub-commission and provisional joint military teams to carry out its tasks concerning captured and detained Vietnamese civilian personnel.

(d) The Four-Party Joint Military Commissions shall send joint military teams to observe the return of persons mentioned in Articles 1 and 2 of this Protocol at each place in Viet-Nam where such persons are being returned, and at the last detention places from which these persons will be taken to the places of return. The Two-Party Joint Military Commission shall send joint military teams to observe the return of Vietnamese civilian personnel captured and detained at each place in South Viet-Nam where such persons are being returned, and at the last detention places from which these persons will be taken to the places of return, the examination of lists, and the investigation of violations of the provisions of the above-mentioned Articles.

Source: United States Treaties and Other International Agreements 1974, 24–31.

Complete Text of the Letter from President Richard Nixon to Prime Minister Pham Van Dong, 1 February 1973

On 1 February 1973 in Paris, France, U.S. Colonel Georges Guay hand-delivered a copy of what became known as "the secret letter" from President Nixon to the prime minister of North Vietnam. In exchange, Guay received a list of prisoners being held in Laos. The list included only ten names. In the letter below, President Nixon appears to offer $3.25 billion in reconstruction aid to the North Vietnamese, a fact that was hidden from the American people until 1977 when the letter was declassified because the Vietnamese demanded that the United States keep its word on the matter.

The President wishes to inform the Democratic Republic of Vietnam of the principles which will govern United States participation in the postwar reconstruction of North Vietnam. As indicated in Article 21 of the Agreement on Ending the War and Restoring Peace in Vietnam signed in Paris on Jan. 27, 1973, the United States undertakes this participation in accordance with its traditional policies.

These principles are as follows:

1. The Government of the United States of America will contribute to postwar reconstruction in North Vietnam without any political conditions.

2. Preliminary United States studies indicate that the appropriate programs for the United States contribution to postwar reconstruction will fall in the range of $3.25 billion of grant aid over five years. Other forms of aid will be agreed upon between the two parties. This estimate is subject to revision and to detailed discussion between the Government of the United States and the Government of the Democratic Republic [of] Vietnam.

3. The United States will propose to the Democratic Republic of Vietnam the establishment of a United States–North Vietnamese Joint Economic Commission within 30 days from the date of this message.

4. The function of the commission will be to develop programs for the United States contribution to reconstruction of North Vietnam. This United States contribution will be based upon such factors as:

(a) The needs of North Vietnam arising from the dislocation of war;

(b) The requirements for postwar reconstruction in the agricultural and industrial sectors of North Vietnam's economy.

5. The Joint Economic Commission will have an equal number of representatives from each side. It will agree upon a mechanism to administer the program which will constitute the United States contribution to the reconstruction of North Vietnam. The commission will attempt to complete this agreement within 60 days after its establishment.

6. The two members of the commission will function on the principle of respect for each other's sovereignty, noninterference in each other's internal affairs, equality and mutual benefit. The offices of the commission will be located at a place to be agreed upon by the United States and the Democratic Republic of Vietnam.

7. The United States considers that the implementation of the foregoing principles will prompt economic, trade and other relations between the United States of America and the Democratic Republic of Vietnam and will contribute to insuring a stable and lasting peace in Indochina. These principles accord with the spirit of

Chapter VIII of the Agreement on Ending the War and Restoring the Peace in Vietnam which was signed in Paris on Jan. 27, 1973.

Understanding Regarding Economic Reconstruction Program

It is understood that the recommendations of the Joint Economic Commission mentioned in the President's note to the Prime Minister will be implemented by each member in accordance with its own constitutional provisions.

Note Regarding Other Forms of Aid

In regard to other forms of aid, United States studies indicate that the appropriate programs could fall in the range of $1 billion to $1.5 billion, depending on food and other commodity needs of the Democratic Republic of Vietnam.

Source: U.S. House of Representatives 1979, Appendix 2: 25.

National Defense Authorization Act, Section 569 on Missing Persons

This 1997 revision of Department of Defense regulations replaced the Missing Persons Act dating from the 1940s, which allowed the Department of Defense to automatically declare service members who became missing in hostile territory dead after one year during which no information surfaces indicating an alive status. The new act mandated the formation of an Office of Missing Personnel within the Office of the Secretary of Defense responsible for the investigation and recovery of missing persons, and set regulations that safeguard the rights of the missing to fair status review process.

TITLE 10. ARMED FORCES SUBTITLE A. GENERAL MILITARY LAW

PART II. PERSONNEL
CHAPTER 76. MISSING PERSONS

10 USCS 1501 (1997)

STATUS: CONSULT PUBLIC LAWS CITED BELOW FOR RECENT CHANGES TO THIS DOCUMENT

LEXSEE 105 P.L. 85—See section 599(a)(1), effective upon enactment.

1501. System for accounting for missing persons

(a) Office for missing personnel.

(1) The Secretary of Defense shall establish within the Office of the Secretary of Defense an office to have responsibility for Department of Defense policy relating to missing persons. Subject to the authority, direction, and control of the Secretary of Defense, the responsibilities of the office shall include—

(A) policy, control, and oversight within the Department of Defense of the entire process for investigation and recovery related to missing persons (including matters related to search, rescue, escape, and evasion); and

(B) coordination for the Department of Defense with other departments and agencies of the United States on all matters concerning missing persons.

(2) In carrying out the responsibilities of the office established under this subsection, the head of the office shall be responsible for the coordination for such purposes within the Department of Defense among the military departments, the Joint Staff, and the commanders of the combatant commands.

(3) The office shall establish policies, which shall apply uniformly throughout the Department of Defense, for personnel recovery (including search, rescue, escape, and evasion).

(4) The office shall establish procedures to be followed by Department of Defense boards of inquiry, and by officials reviewing the reports of such boards, under this chapter [10 USCS 1501 et seq.].

(b) Uniform DOD procedures.

(1) The Secretary of Defense shall prescribe procedures, to apply uniformly throughout the Department of Defense, for—

(A) the determination of the status of persons described in subsection (c); and

(B) for the systematic, comprehensive, and timely collection, analysis, review, dissemination, and periodic update of information related to such persons.

(2) Such procedures may provide for the delegation by the Secretary of Defense of any responsibility of the

Secretary under this chapter [10 USCS 1501 et seq.] to the Secretary of a military department.

(3) Such procedures shall be prescribed in a single directive applicable to all elements of the Department of Defense.

(4) As part of such procedures, the Secretary may provide for the extension, on a case-by-case basis, of any time limit specified in section 1502, 1503, or 1504 of this title. Any such extension may not be for a period in excess of the period with respect to which the extension is provided. Subsequent extensions may be provided on the same basis.

(c) Covered persons. Section 1502 of this title applies in the case of any member of the armed forces on active duty who becomes involuntarily absent as a result of a hostile action, or under circumstances suggesting that the involuntary absence is a result of a hostile action, and whose status is undetermined or who is unaccounted for.

(d) Primary next of kin. The individual who is primary next of kin of any person prescribed in subsection (c) may for purposes of this chapter [10 USCS 1501 et seq.] designate another individual to act on behalf of that individual as primary next of kin. The Secretary concerned shall treat an individual so designated as if the individual designated were the primary next of kin for purposes of this chapter [10 USCS 1501 et seq.]. A designation under this subsection may be revoked at any time by the person who made the designation.

(e) Termination of applicability of procedures when missing person is accounted for. The provisions of this chapter [10 USCS 1501 et seq.] relating to boards of inquiry and to the actions by the Secretary concerned on the reports of those boards shall cease to apply in the case of a missing person upon the person becoming accounted for or otherwise being determined to be in a status other than missing.

HISTORY:

(Added Feb. 10, 1996, P.L. 104-106, Div A, Title V, Subtitle F, 569(b)(1), 110 Stat. 336; Sept. 23, 1996, P.L.
104-201, Div A, Title V, Subtitle H, 578(a)(1), 110 Stat. 2536.)
HISTORY; ANCILLARY LAWS AND DIRECTIVES

Amendments:

1996. Act Sept. 23, 1996, in subsec. (c), substituted "applies in the case of any member" for "applies in the case of the following persons:

"(1) Any member," deleted para. (2), which read: "(2) Any civilian employee of the Department of Defense, and any employee of a contractor of the Department of Defense, who serves with or accompanies the armed forces in the field under orders who becomes involuntarily absent as a result of a hostile action, or under circumstances suggesting that the involuntary absence is a result of a hostile action, and whose status is undetermined or who is unaccounted for"; and deleted subsec. (f), which read: "(f) Secretary concerned. In this chapter, the term 'Secretary concerned' includes, in the case of a civilian employee of the Department of Defense or contractor of the Department of Defense, the Secretary of the military department or head of the element of the Department of Defense employing the employee or contracting with the contractor, as the case may be." Other provisions:

Determination of whereabouts and status of missing persons; purpose of chapter, etc. Act Feb. 10, 1996, P.L. 104-106, Div A, Title V, Subtitle F, 569(a), 110 Stat. 336, provides: "The purpose of this section [adding 10 USCS 1501 et seq.], among other things; for full classification, consult USCS Tables volumes] is to ensure that any member of the Armed Forces (and any Department of Defense civilian employee or contractor employee who serves with or accompanies the Armed Forces in the field under orders) who becomes missing or unaccounted for is ultimately accounted for by the United States and, as a general rule, is not declared dead solely because of the passage of time."

Source: United States Code Service 1997.

The POW/MIA Flag

The POW/MIA flag, rendered below, dates from 1971, when, with the approval of the National League of Families of American Prisoners and Missing in Southeast Asia, it was designed and manufactured. In the years following its creation, various activists and groups have altered the flag, at times changing the colors to red, white, and blue or reversing POW/MIA to MIA/POW.

Since its creation, the flag has been flown over the White House on National POW/MIA Recognition Day and was installed in the United States Capitol Rotunda. In 1997, as part of the fiscal year 1998 National Defense Authorization Act, the flag was allowed to be flown at such federal sites as certain military bases, selected war memorials, and the United States Post Office. This legislation also requires the flag to be flown at federal buildings on patriot holidays, such as Armed Forces Day, Memorial Day, Flag Day, Independence Day, Veterans Day, and National POW/MIA Recognition Day.

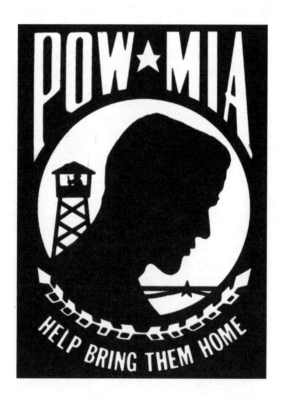

References

Clinton, William. "Proclamation 6818: National POW/MIA Recognition Day, 1995." 1995a. *Weekly Compilation of Presidential Documents* 31, no. 35 (4 September). Office of the Federal Register, National Archives and Records Services, General, General Services Administration. Washington, DC: U.S. Government Printing Office.

———. "U.S. Normalizes Diplomatic Relations with Vietnam." 1995b. *U.S. Department of State Dispatch* no. 28 (10 July). Office of Public Communication. Washington, DC: U.S. Government Printing Office.

Dole, Robert. 1995. "Relations with Vietnam." *Congressional Record* 141, no. 110 (10 July): S9626–S9628.

Griffiths, Ann Mills. 1994. "Position on United States Relations with Vietnam in the Context on POW/MIA Progress." *Congressional Record* 140, no. 3 (27 January), S211.

Hackworth, David H., and Julie Sherman. 1989. *About Face.* New York: Simon and Schuster.

Kutler, Stanley I., ed. 1996. *Encyclopedia of the Vietnam War.* New York: Charles Scribner's Sons.

Lord, Winston. 1994. "U.S.-Vietnam POW/MIA Progress: Lifting the Embargo."*U.S. Department of State Dispatch* 5, no. 9 (28 February). Office of Public Communication. Washington, DC: U.S. Government Printing Office.

McDaniel, Eugene, and James Johnson. 1975. *Scars and Stripes.* Philadelphia: A. J. Holman.

National League of Families of American Prisoners and Missing in Southeast Asia. 1997. *Factsheet* (25 November).

Office of Assistant Secretary of Defense (Public Affairs). 1994. News Release No. 028-94 (24 January): 3.

Olson, James S. 1988. *Dictionary of the Vietnam War.* Westport, CT: Greenwood Press.

Rowan, Stephan A. 1973. *They Wouldn't Let Us Die: Prisoners of War Tell Their Story.* Middle Village, NY: Jonathan David Publishers.

United States Code Service, §1501 (1997) Title 10. Armed Forces, Subtitle A. General Military Law, Part II. Personnel. Chapter 76, Missing Persons, LEXIS Law Publishing.

United States Treaties and Other International Agreements. 1956. Vol. 6, part 3. Washington, DC: U.S. Government Printing Office.

United States Treaties and Other International Agreements. 1974. Vol. 24, part 1. Washington, DC: U.S. Government Printing Office.

U.S. Department of Defense. 1998. "1997 Identified Remains."

POW/MIA Newsletter (spring): n.p. http://lcweb2.loc.gov/pow.

———. 1990. *POW-MIA Factbook.* Washington, DC: U.S. Government Printing Office.

———. "Remains of U.S. Servicemen from Southeast Asia Identified." POW/MIA Weekly Update (1 April). http://lcweb2.loc.gov/pow.

———. 1979. *Selected Manpower Statistics,* Headquarters for Information and Operations and Reports, Washington, DC: Directorate for U.S. Government Printing Office.

———. 1985. *U.S. Casualties in Southeast Asia: Statistics as of April 30, 1985.* Washington Headquarters Services Directorate for Information Operations and Reports. Washington, DC: U.S. Government Printing Office.

U.S. House of Representatives. 1996a. *Accounting for U.S. POW/MIAs in Southeast Asia.* Hearing before the Military Personnel Subcommittee, Committee on National Security, 104th Congress, first session, 29 June 1995. Washington, DC: U.S. Government Printing Office.

———. 1979. *Aid to North Vietnam.* Hearing before the Subcommittee on International Relations, 95th Congress, first session. Washington, DC: U.S. Government Printing Office.

———. 1996b. *Department of Defense's Comprehensive Review of POW/MIA Cases.* Hearings before the Military Personnel Subcommittee, Committee on National Security, 104th Congress, first session, 20, 30 November 1995. Washington, DC: U.S. Government Printing Office.

———. 1996c. *United States and Vietnamese Government Knowledge and Accountability for U.S. POW/MIAs.* Hearing before the Military Personnel Subcommittee, Committee on National Security, 104th Congress, first session, 14 November. Washington, DC: U.S. Government Printing Office.

U.S. Senate. Committee on Veterans Affairs. 1980. *Study of Former Prisoners of War.* Washington, DC: U.S. Government Printing Office.

Veith, George J. 1998. *Code-Name Bright Light: The Untold Story of U.S. POW Rescue Efforts during the Vietnam War.* New York: Free Press.

Organizations 5

This chapter describes a sampling of organizations, listed alphabetically, that in some way deal with the POW/MIA issue. They may be veterans organizations, or more particularly groups for former POWs, MIA family groups, or family/activist groups that allow the membership of nonfamily members.

Advocacy and Intelligence Index
for Prisoners of War Missing in Action
1220 Locust Ave., Bohemia
Long Island, NY 11716-2169
(516) 567-9057
Fax: (516) 244-7097
E-mail: aiipowmiai@aol.com
Web site: http://www.aiipowmia.com

Works to affect government policy on POWs and information pertaining to them. Disseminates information useful to POW/MIA activists through its Web site and E-mail list server.

American Defense Institute
1055 N. Fairfax St., 2nd Floor
Alexandria, VA 22314
(703) 519-7000

Founded in 1983 by former POW Eugene "Red" McDaniel, the American Defense Institute (ADI) is the educational arm of the

American Defense Foundation, also founded by McDaniel. Both organizations promote the "value of freedom and our responsibility for protecting it with a strong national defense." The institute funds fellowships in national security studies, internships in national defense studies, and an adjunct scholar program. It also conducts a series of seminars on national security issues and runs a nonpartisan voter registration program among military personnel. Its POW awareness campaign focuses on educating the public about efforts to account for MIAs through a national speakers bureau, public relations events, and press releases.

Publications: Publications include *ADI News,* a quarterly membership newsletter about ADI activities; *ADI Briefing,* a monthly series of position papers; and the *POW/MIA News,* a bimonthly report that highlights news on the POW issue.

American Ex-Prisoners of War, Inc.
c/o Clydie J. Morgan
3201 E. Pioneer Pkwy., No. 40
Arlington, TX 76010-5396
(817) 649-2979
Fax: (817) 649-0109
E-mail: pow@flash.net

Membership is open to all persons on active duty or retired from the armed services of the United States who were held prisoners of war by an enemy of the United States because of such service, U.S. citizens who were imprisoned because of their citizenship, and U.S. citizens who were taken prisoners by an enemy of the United States while honorably serving an ally of the United States. The following relatives may hold a full membership in the organization: spouse, parents, stepparents, brothers, sisters, children, and heirs-in-law as defined in the Washington Revised Statutes as amended. Its purpose is "to encourage fraternity for the common good, foster patriotism and loyalty, assist Ex-Prisoners of War, Civilian Internees, and their widows and orphans, maintain allegiance to the United States of America, preserve and defend the United States from all of her enemies, [and] maintain historical records." The organization holds national and chapter conventions, provides support to former POWs and their families, supports legislation concerning POWs and related topics, and disseminates medical research and information. Its services include volunteers at Veterans Administration hospitals and national officers who help ex-POWs and family members file benefits claims through the Veterans Benefits Administration. The group was founded in 1942 and has 33,000 members.

Publication: A monthly bulletin, *EX-POW BULLETIN,* which contains information on reunions, activities, medical research of interest to former POWs, book reviews, personal recollection, and obituaries.

American Ex-Prisoners of War, Inc.—Washington, D.C., Office
c/o Bill Rolen
National Capitol Office, Rm. 1118
1120 Vermont Ave. NW
Washington, DC 20421-1111
(202) 418-4258
Fax: (202) 463-9769

American Ex-Prisoners of War, Inc.—Regional Chapters

Eastern Zone
c/o Zack Roberts
374 Rolling Rock
Mountainside, NJ 07092
(908) 232-2990
Fax: (908) 232-6450

Central Zone
c/o Harley J. Coon
2439 Lantz Rd.
Beavercreek, OH 45434
(512) 426-5105
Fax: (513) 426-8415

Western Zone
c/o Jack E. Jones
1318 98th NE
Bellevue, WA 98004
(206) 454-1972
Fax: (206) 688-7686

East Central Region
c/o Frank A. Kravetz
105 Elizabeth Ave.
E. Pittsburgh, PA 15112
(412) 824-2674

Northeast Region
c/o Paul E. Butler
1355 Wantagh
Wantagh, NY 11793
(516) 221-6903

Southeast Region
c/o Buel Knight
16695 Hwy. 69 N
Northport, AL 35476
(205) 339-4675

North Central Region
c/o John Klumpp
2000 Diana Dr.
Lincoln, NE 68506
(402) 488-5964

Mid-Central Region
c/o Peter L. Choma
1418 Mill Rd.
Trenton, IL 62293-2902
(618) 224-7472

South Central Region
c/o Harland J. Hendrix

Rt. 2, Box 261-A
Meeker, OK 74855
(405) 275-5962

Northwest Region
c/o Kenneth Viles
2710 Weigel Ave.
Vancouver, WA 98660
(360) 695-8708
Fax: (360) 992-0842

Southwest Region
c/o Robert C. Clark
107 S. Los Robles Ave.
Pasadena, CA 91101
(818) 577-0421
Fax: (818) 796-5424

Department of Alabama
c/o Cmdr. Ida Davis
P.O. Box 40253
Tuscaloosa, AL 35404
(205) 553-3832

Department of Arizona
c/o Cmdr. Winona (Tootie)
    Ney
575 Paseo Reforma
Rio Rico, AZ 85648
(520) 281-8771

Department of California
c/o Cmdr. Raymond Merritt
688 Saddleback Dr.
Marysville, CA 95901
(916) 749-8324

Department of Colorado
c/o Cmdr. Clayton A. Nattier
2556 S. Eaton Pl.
Lakewood, CO 80227
(303) 988-2579

Department of Florida
c/o Cmdr. Joseph R. Garber
544 Oak Island Cr.
Plant City, FL 33565
(813) 754-3010

Department of Georgia
c/o Cmdr. Bernard Porter
3315 Sandpiper Ct.
Augusta, GA 30907
(706) 863-8006

Department of Illinois
c/o Cmdr. Diane Rose
1335 Daffodil Ln.
Highland, IL 62249
(618) 654-7780

Department of Indiana
c/o Cmdr. Robert Waldrop
2711 Northaven Ct.
Fort Wayne, IN 46825
(219) 483-0886

Department of Iowa
c/o Cmdr. Bruce A. Yungclas
1328 Grand St.
Webster City, IA 50595-2647
(515) 832-3589

Department of Kentucky
c/o Cmdr. Dudley Riley
412 Hickory St.
Dawson Springs, KY
    42408-1612
(502) 797-4147

Department of Louisiana
c/o Cmdr. Warren W. Duncan
8771 Cottage Ave.
Baton Rouge, LA 70806
(504) 924-2460

Department of Maryland
c/o Cmdr. John Holmes
  Meyers
7536 Brightwater Beach Rd.
Glen Burnie, MD 21061
(301) 437-4788

Department of Massachusetts
c/o Cmdr. Samuel Palter
12 Schirmer Rd.
W. Roxbury, MA 02132
(617) 325-7127

Department of Minnesota
c/o Cmdr. Earle E.
  Bombardier
506 W. Langen St.
Battle Lake, MN 56516
(218) 864-8917

Department of Mississippi
c/o Cmdr. Carey Ashcraft
67 Shores Dr.
Clinton, MS 39056
(601) 924-7286

Department of Missouri
c/o Cmdr. Owen B. Pickle
2180 Canterbury Dr.
Florissant, MO 63033-1204
(314) 839-3409

Department of Nebraska
c/o Cmdr. Ben Comstock
2806 Nottingham Dr.
Bellevue, NE 68123
(402) 291-6338

Department of New Jersey
c/o Cmdr. Dick McAuley
731 1st Ave.
Dunellen, NJ 08812
(908) 968-0224

Department of New Mexico
c/o Cmdr. John J. Mirabal
P.O. Box 236
Tularosa, NM 88352
(505) 585-9050

Department of New York
c/o Cmdr. Edward Curtin
4 American Ave.
Coram, Long Island, NY 11727
(516) 698-4274

**Department of North
  Carolina**
c/o Cmdr. F. Paul Dallas
916 Bingham Dr.
Fayetteville, NC 28304
(910) 867-2775

Department of North Dakota
c/o Cmdr. Arnold Postovit
P.O. Box 5
Tioga, ND 58852
(701) 664-2144

Department of Ohio
c/o Cmdr. Maynard (Doc)
  Unger
1316 Sloane Ave.
Lakewood, OH 44107
(216) 521-1889

Department of Oklahoma
c/o Cmdr. Robert Boulware
4900 N.W. 26th
Oklahoma City, OK 73127
(405) 942-4900

Department of Oregon
c/o Cmdr. Mike Ryan
339 S. 7th St.
Lebanon, OR 97355-2212
(541) 258-5732

Department of Pennsylvania
c/o Cmdr. Joseph Leichte
903 Valley St.
Enola, PA 17025
(717) 732-4948

**Department of South
Carolina**
c/o Cmdr. Robert Lammey
936 Bel Aire Dr.
Rock Hill, SC 29732
(803) 327-3956

Department of South Dakota
c/o Cmdr. Clair Headrick
703 W. Elm
Flandreau, SD 57028
(605) 997-2617

Department of Tennessee
c/o Cmdr. Ivo Dillon
2802 Edwards Dr.
Nashville, TN 37211-2410
(615) 833-4205

Department of Texas
c/o Cmdr. Michael G. Gesino,
  Sr.

3121 Brincrest
Farmers Branch, TX 75234
(214) 247-2925

Department of Virginia
c/o Cmdr. George Juskalian
6604 Ashmere Ln.
Centreville, VA 20120
(703) 222-0066

Department of Washington
c/o Cmdr. Jerome V. Gleesing
E. 13421 25th
Spokane, WA 99216-0489
(509) 928-5458

Department of West Virginia
c/o Cmdr. Harold L.
  Lumadue
209 17th St.
Dunbar, WV 25064
(304) 768-7851

Department of Wisconsin
c/o Cmdr. Howard M. Jones
1005 King's Way
Nekossa, WI 54457
(715) 325-5955

American GI Forum of United States
1315 Bright St.
Corpus Christi, TX 78405
(512) 882-2123

Founded in 1948, this 20,000-member organization serves veterans of the U.S. armed forces, primarily of Mexican origin. Its purpose is to promote the principles of democracy and equal opportunity. It conducts programs on education, business development, and POWs and veterans affairs.

American Legion
c/o Public Relations Division
700 N. Pennsylvania St.

Indianapolis, IN 46204
(317) 630-1200
Fax: (317) 630-1368
E-mail: tal@iquest.net
Web site: http://www.legion.org

This organization, which promotes patriotism and aids veterans, is open to honorably discharged wartime veterans, both male and female, of the U.S. armed forces. The legion maintains a museum and library, conducts educational programs, and promotes veterans' issues through the legislative process. Founded in 1919, the legion has more than 2.8 million members.

Publications: Monthly magazine, *American Legion Magazine*; biweekly newspaper, *Dispatch.*

American Legion—State

American Legion of Alabama
P.O. Box 1069
Montgomery, AL 36101
(205) 262-6638

American Legion of Alaska
519 W. 8th Ave., Ste. 208
Anchorage, AK 99501
(907) 278-8598

American Legion of Arizona
4701 N. 19th Ave., Ste. 200
Phoenix, AZ 85015
(602) 264-7706

American Legion of Arkansas
P.O. Box 3280
Little Rock, AR 72203
(501) 375-4236

**American Legion of
California**
117 Veterans War Memorial
Bldg.
San Francisco, CA 94102-4587
(415) 431-2400

American Legion of Colorado
3003 Tejon St.
Denver, CO 80211
(303) 477-1655

**American Legion of
Connecticut**
P.O. Box 208
Rocky Hill, CT 06067
(203) 721-5921

**American Legion of
Delaware**
P.O. Box 5696
Wilmington, DE 19808
(302) 998-9448

**American Legion of the
District of Columbia**
3408 Wisconsin Ave. NW,
Ste. 212
Washington, DC 20016
(202) 362-9151

American Legion of Florida
P.O. Box 547936

Orlando, FL 32854-7936
(407) 295-2631

American Legion of Georgia
3282 E. Main St.
College Park, GA 30337
(404) 767-0007

American Legion of Hawaii
612 McCully St.
Honolulu, HI 96826
(808) 946-6383

American Legion of Idaho
901 Warren St.
Boise, ID 83706
(208) 342-7061

American Legion of Illinois
P.O. Box 2910
Bloomington, IL 61702
(309) 663-0631

American Legion of Indiana
777 N. Meridian St.
Indianapolis, IN 46204
(317) 630-1263

American Legion of Iowa
720 Lyon St.
Des Moines, IA 50309
(515) 282-5068

American Legion of Kansas
1314 S.W. Topeka Blvd.
Topeka, KS 66612
(913) 232-9315

**American Legion of
   Kentucky**
P.O. Box 2123
Louisville, KY 40201
(502) 587-1414

**American Legion of
   Louisiana**
P.O. Box 1431
Baton Rouge, LA 70821
(504) 923-1945

American Legion of Maine
P.O. Box 900
Waterville, ME 04901
(207) 873-3229

**American Legion of
   Maryland**
War Memorial Bldg., Rm. E
101 N. Gay St.
Baltimore, MD 21202
(401) 752-1405

**American Legion of
   Massachusetts**
State House 546-2
24 Beacon St.
Boston, MA 02133
(617) 727-2966

**American Legion of
   Michigan**
212 N. Verlinden Ave.
Lansing, MI 48915
(517) 371-4720

**American Legion of
   Minnesota**
State Veterans Service Bldg.
St. Paul, MN 55155
(612) 291-1800

**American Legion of
   Mississippi**
P.O. Box 688
Jackson, MS 39205
(601) 352-4986

American Legion of Missouri
P.O. Box 179
Jefferson City, MO 65102
(573) 893-2353

American Legion of Montana
P.O. Box 6075
Helena, MT 59604
(406) 442-5260

American Legion of Nebraska
P.O. Box 5205
Lincoln, NE 68505
(402) 464-6338

American Legion of Nevada
737 Veterans Memorial Dr.
Las Vegas, NV 89101
(702) 382-2353

American Legion of New
  Hampshire
State House Annex
25 Capitol St., Rm. 431
Concord, NH 03301-6312
(603) 271-5338

American Legion of New
  Jersey
101 S. Broad St.
Trenton, NJ 08608
(609) 695-5418

American Legion of New
  Mexico
1215 Mountain Rd. NE
Albuquerque, NM 87102
(505) 247-0400

American Legion of New York
112 State St., Ste. 400
Albany, NY 12207
(518) 463-2215

American Legion of North
  Carolina
P.O. Box 26657
Raleigh, NC 27611
(919) 832-7506

American Legion of North
  Dakota
P.O. Box 2666
Fargo, ND 58108-2666
(701) 293-3120

American Legion of Ohio
P.O. Box 14348
Columbus, OH 43214
(614) 268-7072

American Legion of
  Oklahoma
P.O. Box 53037
Oklahoma City, OK 73152
(405) 525-3511

American Legion of Oregon
P.O. Box 1730
Wilsonville, OR 97070
(503) 685-5006

American Legion of
  Pennsylvania
P.O. Box 2324
Harrisburg, PA 17105-2324
(717) 730-9100

American Legion of Rhode
  Island
83 Park St., Rm. 403
Providence, RI 02903
(401) 421-7390

American Legion of South
  Carolina
P.O. Box 11355

Columbia, SC 29211
(803) 799-1992

American Legion of South Dakota
P.O. Box 67
Watertown, SD 57201-0067
(605) 886-3604

American Legion of Tennessee
215 8th Ave. N
Nashville, TN 37203
(615) 254-0568

American Legion of Texas
P.O. Box 789
Austin, TX 78767
(512) 472-4138

American Legion of Utah
B-61 State Capitol Bldg.
Salt Lake City, UT 84114
(801) 538-1013

American Legion of Vermont
P.O. Box 396
Montpelier, VT 05602
(802) 223-7131

American Legion of Virginia
P.O. Box 11025
Richmond, VA 23230
(804) 353-6606

American Legion of Washington
P.O. Box 3917
Lacey, WA 98503-3917
(206) 491-4373

American Legion of West Virginia
P.O. Box 3191
Charleston, WV 25332
(304) 343-7591

American Legion of Wisconsin
812 E. State St.
Milwaukee, WI 53202
(414) 271-1940

American Legion of Wyoming
P.O. Box 545
Cheyenne, WY 82003
(307) 634-3035

American POW-MIA Coalition
c/o Charlotte Alden-White
6301 W. 81st St.
Overland Park, KS 66204-3902
(913) 649-8051
Fax: (913) 649-8051

Founded in 1992, this 290-member organization works to affect government policy on POW/MIAs and information pertaining to them. With membership open to individuals and organizations, the coalition conducts public awareness events about military personnel listed as missing in action, supports families of MIAs, attempts to change government policy on the MIA issue, offers educational programs, and maintains an archive of information on the issue, including the biographies of some 2,600 POWs.

Publication: Periodic newsletter.

American Veterans of World War II, Korea,
and Vietnam (AMVETS)
4647 Forbes Blvd.
Lanham, MD 20706-4380
(301) 459-9600
Fax: (301) 459-7924
E-mail: amvets@amvets.org

Veterans of World War II, the Korean War, and Vietnam make up this organization (known as AMVETS), which provides a variety of services. Founded in 1944, AMVETS has 176,000 members. Among its many committees are those that deal with employment, health issues, homelessness, and scholarships. At the national level a POW/MIA affairs committee "reviews, considers information about, and promotes the safe repatriation of POWs and attempts to keep the public aware of the POW/MIA issue." This committee makes recommendations about legislative action and helps former POWs lead productive lives.

Publication: A quarterly magazine, *The National AMVET.*

Canadian POW*MIA Information Centre—
Norman A. Todd Chapter
41 Laurier Ave.
Milton, Ontario
L9T 4T1 Canada
(905) 875-0658
Fax: (905) 875-4232
E-mail: mgillhoo@netrover.com
Web site: http://www.ipsystems.com/powmia

In order to promote "public awareness on the plight of POW/MIAs, and to bring about a termination to the shameful practice of our governments to abandon our men and women to fates unknown at the hands of our enemies," this organization engages in a number of activities. Among them are speaking at high schools, colleges, and universities; giving families of POWs assistance and support; and displaying a POW/MIA awareness booth at trade shows and other events. Founded in 1989, the group has 153 members in five chapters in southern Ontario. Membership is free.

Publications: Monthly newsletter, pamphlets, documents, and a Web site.

**Canadian POW*MIA Information Centre—
Austin M. Corbiere Chapter**
18 Rule St.
Garson, Ontario
P3L 1A5 Canada
E-mail: wilson@cyberbeach.net

**Canadian POW*MIA Information Centre—
Jonathan P. Kmetyk Chapter**
c/o 190 Borrows St.
Thornhill, Ontario
L4J 2W8 Canada
(905) 738-0104

**Canadian POW*MIA Information Centre—
Michael (Bat) Masterson Chapter**
RR #2
Blyth, Ontario
N0M 1H0 Canada
(519) 523-9427
E-mail: canadianrose@geocities.com
Web site: http://www.geocities.com/SouthBeach/Marina/9680

**Canadian POW*MIA Information Centre—
William Butler Chapter**
100 Main St. W
Otterville, Ontario
N0T 1R0 Canada
(519) 879-6205

Chosin Few
c/o Win K. Scott
30 Mountain Spring Rd.
Waynesville, NC 28786-9781
(828) 452-7124
Fax: (828) 452-7124
E-mail: chosinhq@aol.com
Web site: http://www.lava.net.wardsway

This organization is open to veterans of the United States, the Republic of Korea, Great Britain, and Australia who fought around the Chosin Reservoir in North Korea in November–December 1950. Its members work to account for Americans thought to still be prisoners of war in North Korea and repatriate the remains of American MIAs. The group was founded in 1983 and has 5,600 members.

Publication: A bimonthly magazine, *Chosin Few Digest,* which includes recollections, articles about present-day Korea, and chapter news.

Chosin Few—Regional Groups

Northeast Region
c/o Peter V. Brask
22 Sand Point Rd.
Kennebunkport, ME 04046
(207) 967-2026

c/o Harry W. Hogan
204 Ave. O
West Wildwood, NJ 08260
(609) 729-5312

c/o Francis H. "Timmy"
   Killeen
P.O. Box 295
Rocky Point, NY 11778-0295

Southeast Region
c/o Ernie P. Bond
3136 Brunswick Cr.
Palm Harbor, FL 34684-2305

c/o Jack Hessman
12916 Penrose St.
Rockville, MD 20853
(301) 933-6529

c/o Col. Robert E. Parrott
8818 Mt. Vernon Hwy.
Alexandria, VA 22309-2219
(703) 360-9203

Midwest Region
c/o Col. Robert A. Henderson
807 S. Maple St.
Urbana, IL 61801-4208
(217) 367-9974

c/o Richard E. Oly
2386 Huntingdale Ln.
Oviedo, FL 32765-3430
(407) 366-3430

c/o Richard W. Danielson
4575 Westview Dr.
North Olmstead, OH 44070
(216) 777-9677

Western Region
c/o Manert H. Kennedy
1515 Elder
Boulder, CO 80304-2629
(303) 444-3803

c/o Kenneth F. Santor
25 Southridge Dr.
Reno, NV 89509-3254
(702) 322-1975

West Coast Region
c/o Thomas G. Green
10105 S.E. 96th Ave.
Portland, OR 97266
(503) 777-2947

c/o Lt. Col. Thomas Kalus
98-1927 Wilou St.
Alea, HI 96701-1663
(808) 486-5004

c/o Howard R. Mason
1730 242nd Pl.
Lomita, CA 90717-1309
(310) 326-6729

Committee for the Defense of Political Prisoners in Vietnam
c/o Nguyen Huu Hieu
15701 Prosperity Dr.
Haymarket, VA 22069
(703) 754-0483

Made up of former prisoners, their families and friends, students, refugees, and other concerned individuals, this organization works to call attention to the plight of political prisoners in the Socialist Republic of Vietnam. Founded in 1978, it seeks better human rights for those in Vietnam detained in reeducation camps, urging Vietnam to adhere to United Nations principles of human rights. It conducts research via fact-finding missions, documentation, and interviews with former prisoners and their families and disseminates this information through press conferences and other methods. It also solicits funds to buy medicine to send to prisoners.

Publication: A monthly magazine, *Thoi Tap,* which includes nonfiction, fiction, poetry, reviews, and translations.

Heart of Illinois POW/MIA Association
P.O. Box 1193
Pekin, IL 61555
(309) 353-5564

Founded in 1986 with the goal of bringing home live POWs known to have been left behind in Vietnam, achieving the fullest possible accounting of the missing, and preventing men from being left behind in the future, this organization conducts public awareness and fund-raising activities. With 2,700 members, it also assists individual families of POWs/MIAs in their efforts to obtain a full accounting.

Publication: A quarterly newsletter.

Indiana State Veterans Coalition
c/o Dick Forrey
3210 Chelsea Ct.
Kokomo, IN 46902
(765) 455-0048
Fax: (765) 453-4366

Open to honorably discharged wartime veterans, both male and female, of the U.S. armed services, this 2,362-member organization works to better the lives of veterans and see the return of live POWs and account for the missing from U.S. wars. Founded in 1991, the coalition assists in Veterans Administration hospitals, networks with other POW/MIA groups, supports research activities conducted out of Thailand, and provides speakers upon request.

Jewish War Veterans of the U.S.A.
National Ladies Auxiliary
c/o Rita A. Panitz
1811 R St. NW
Washington, DC 20009
(202) 667-9061

This 15,000-member organization is made up of the sisters, wives, mothers, daughters, widows, and descendants of Jewish veterans of U.S. wars. Founded in 1928, it works on many fronts to aid veterans, particularly the homeless. Its MIA/POW Red Ribbon Campaign focuses on educating the public about efforts to account for missing servicemen.

Korean/Cold War Family Association of the Missing
128 Beaver Run
Coppell, TX 75019-4849
(972) 471-0246 (evenings only)

Full membership in this 900-member organization is open to any family member of a Korean War MIA; nonvoting membership is open to any interested person. The goal of this association, which was founded in 1993, is to aid the families of Korean War MIAs to account for the missing through the return of remains and acquisition of pertinent information. To this end, it conducts archival research, disseminates research results to family members, makes public awareness efforts aimed at national legislators, and supports relevant legislation and international activities.

Publication: A newsletter.

Korean War Veterans National Museum and Library
c/o 700 S. Main
Tuscola, IL 61953 (217) 253-2535; (217) 253-4620
E-mail: dcmuseum@net66.com
Web site: http://www.ameritech.net/users/decker/museum. htm

This historical preservation agency's specific purpose is to establish a national museum and library for Korean War veterans, which will preserve a record of U.S. participation in the Korean War; educate the public about the Korean War; and promote good relations among Korean War veterans of all countries and the people of South Korea. In addition, the museum and library will collect, research, house, and interpret Korean War information and artifacts.

Live POW Lobby
c/o Mike Van Atta
7035 Ingalls Ct.
Arvada, CO 80003-3724
(303) 467-6847

This organization aims to publicize the plight of American servicemen who are POW/MIAs in Southeast Asia and seeks their return. It educates the public, conducts research, and supports relevant legislation on the POW/MIA issue.

Publication: A newsletter, *Insider.*

**The National Alliance of Families for the
Return of America's Missing Servicemen**
5021 133rd Ave. NE
P.O. Box 40327
Bellevue, WA 98004-0327
(425) 881-1499
Fax: (425) 462-1586
Web site: http://www.nationalalliance.org/

In order to account for the missing from World War II, the Korean War, the Cold War, and Vietnam, this organization educates the public about POW/MIA issues and supports relevant legislation.

Publication: Bits and Pieces, a newsletter at its Web site.

**National League of Families of American
Prisoners and Missing in Southeast Asia**
1001 Connecticut Ave. NW, Ste. 919
Washington, DC 20036-5504
(202) 223-6846
(202) 659-0133 (24-Hour Update Line)
Fax: (202) 785-9410
E-mail: NatlLeague@aol.com *or* 76142.611@compuserve.com
Web site: under development

Full members in this organization are the wives, children, parents, and other close relatives of Americans who are listed as POW, MIA, KIA/BNR, and returned U.S. POWs of Vietnam. Associate members include extended relatives of POW/MIAs and concerned citizens. The league's sole purpose is to obtain the release of all prisoners and the fullest possible accounting for the missing and repatriation of all recoverable remains of those who died serving our nation during the Vietnam war. Its activities include lobbying Congress, making public awareness efforts, and fund-raising. The group was founded in 1970 and has 3,850 members. It convenes an annual convention in Washington, D.C.

Publications: National League of Families of American Prisoners and Missing in Southeast Asia; bimonthly newsletter; factsheets.

National Vietnam POW Strike Force
c/o Joe Jordan
2615 Waugh Drive
Suite 217
Houston, TX 77006-2799
(713) 680-3181
Fax: (713) 680-3185

This advocacy group is made up of Vietnam veterans, MIA families, and interested citizens. Networking with other like organizations, its goal is to gather and disseminate information on the POW/MIA issue.

Northeast POW/MIA Network
50 Ferris St.
St. Albans, VT 05478

This organization educates the public about imposter POWs by researching and distributing warnings concerning phony POWs and other veterans. Send complaints or inquiries to the P.O.W. Network listed below.

Operation Just Cause
P.O. Box 264
Stockholm, NJ 07460
Web site: http://whitetail.nji.com/~gfallon/

The purpose of Operation Just Cause is to "bring the POW/MIA issue up close and personal to Congress" through lobbying by individuals on behalf of particular MIAs. The organization matches individuals with MIAs and provides a biography of the MIA and a sample letter for the individual to use to write his or her senators and congressmen.

P.O.W. Network
P.O. Box 68
Skidmore, MO 64487-0068
(660) 928-3304
Fax: (660) 928-3303
Bulletin board system: (660) 928-3305
E-mail: Chuck/Mary Schantag: pownet@asde.com
or Dan/Deb Stock: powpxdds@cloudnet.com
Web site: http://www.msc-net.com//pownet/pownet.htm

Founded in 1987, this network disseminates information on POWs and MIAs. "We feel that the instant availability of news allows activists from around the country to act with a unity of purpose, and a uniformity of information. Lack of information can no longer be an excuse," explains Schantag. The network provides a computerized bulletin board service, Web site, and compilation of archival documents, including biographies of prisoners of war. It creates specialty software products, such as P.O.W. Biography Database, The Wall, Operation Smoking Gun, and The Korean Conflict Casualty List and will reprint for individuals articles from its substantial archive, which is indexed at the network Web site.

POW/MIA FOIA Litigation Account
c/o Roger Hall
8715 1st Ave., No. 827
Silver Spring, MD 20910
(301) 585-3361
E-mail: rhall8715@aol.com

Roger Hall, an archives researcher experienced in retrieval of POW/MIA data, solicits funds through this escrow account to be used to finance legal activities involved in researching POW and MIA cases. When a request for information made under the Freedom of Information Act is denied, legal action can be used to enforce the executive orders to declassify documents relevant to POWs and MIAs.

POW's & MIA's Project Interstate
4210 POW & MIA Memorial Dr.
St. Cloud, FL 34772-8142

(407) 892-9006; (407) 957-MIAS

E-mail: powmia19@gdi.net

This organization wants to create an interconnecting U.S. interstate highway bearing the honorary name POW & MIA Interstate. Thus, it lobbies to pass Senate Bill S-1042 to name an interconnecting U.S. interstate highway in honor of POWs and MIAs.

Red Badge of Courage, Inc.
c/o Adrian Fisch
1023 5th Ave. S
Saint James, MN 56081-2123
(507) 375-5435
Fax: (507) 375-4142

This organization supports efforts to gain a full accounting of all military personnel missing in action during the Korean and Vietnam wars. It conducts public awareness efforts, provides public speakers, and raises funds. Founded in 1970, it maintains a large archive of relevant documents and fulfills requests for information.

Rolling Thunder, Inc.
c/o Artie Muller
National Chapter 1
P.O. Box 216
Neshanic Station, NJ 08853
(908) 369-5439; (908) 904-0373

This organization is made up of veterans and interested citizens, many of whom own motorcycles. In an annual Memorial Day event dating from 1987, the riders parade through Washington, D.C., for a ceremony at the Vietnam Veterans Memorial. "It is a demonstration for the inaction on the part of the U.S. government towards the POW-MIA issue. We ride for those who can't speak for themselves, the POW-MIAs. Our mission is to bring home all the live American POWs, to bring home all remains and to protect the future veterans in the United States armed forces." This organization also operates an escrow account and accepts donations to underwrite the legal fees of Robert Garwood, who is legally challenging his court-martial, in which he was judged to have collaborated with the North Vietnamese during his captivity. Garwood and his supporters maintain that he was a prisoner of war—not a collaborator—and should be entitled to his back pay and benefits, like other former POWs.

S.E.A.R.C.H. Inc.
162 Pine Valley Ln.
Cuba, MO 65453
(573) 885-2778
E-mail: perry@fidnet.com
Web site: http://www.iwc.com/POW-Search/

"Under our charter, we have but one agenda. That is to do what we can to secure the release of Americans held against their will in Communist hands." This organization funds research teams that travel to Southeast Asia to develop intelligence on POWs and MIAs. Information is then disseminated through its Web site.

Support POW/MIA of Austin
c/o Sue Sullivan
3810 Steck Ave.
Austin, TX 78759
(512) 343-5465

To ensure an accounting of American servicemen listed as prisoner of war or missing in action from the Vietnam war, members of this organization give speeches to educate the public that live POWs and MIAs must be accounted for. Founded in 1968, it numbers between 75 and 100 members.

Task Force Omega
14043 N. 64th Dr.
Glendale, AZ 85306
(602) 979-5651
Fax: (602) 979-5651
E-mail: tfoinc@starlink.com

In order "to return all POWs, both alive and dead," Task Force Omega offers educational programs, conducts symposia, and conducts research on the issue and on individual cases from all U.S. wars. Founded in 1983, the group maintains extensive archives on MIAs, including biographies.

Task Force Omega—State Groups

Task Force Omega of Alabama
c/o Randy Hodge
133 Worthington Cr.
New Market, AL 35761
(205) 379-2992

Task Force Omega of Central California
c/o Luann Pike
71 Hickory
Lemoor, CA 93245

(209) 924-3661
Fax: (209) 924-8874

**Task Force Omega of
Colorado**
c/o Lea Dickinson
6230 Tuckerman Ln.
Colorado Springs, CO 80918
(719) 599-7560

Task Force Omega of Florida
c/o James "Grits" and Barbara
Dundon
529 Coppitt Dr. S
Orange Park, FL 32073
(904) 278-8649
Fax: (904) 278-8649

Task Force Omega of Indiana
c/o Ann and Jack Sanders
808 Modrell Blvd.
Elkhart, IN 46514
(219) 262-0760
Fax: (219) 262-2312

**Task Force Omega of
Kentucky**
c/o Danny "Greasy" and
Carol Belcher
P.O. Box 44
Preston, KY 40366
(606) 674-6799

**Task Force Omega of
Louisiana**
c/o Dennis Phillips
3605 Breville St.
Monroe, LA 71203
(318) 343-6619

**Task Force Omega of
Michigan**
c/o Gene Hensell

54351 Parkville Rd.
Mendon, MI 49072
(616) 496-7925

**Task Force Omega of
Nebraska**
c/o Randy Holke
202 S. Howard
Fremont, NE 68025
(402) 727-7970
Fax: (402) 721-6749

**Task Force Omega of New
York**
c/o Darryl D'Agostino
80 La Salle Ave.
Kenmore, NY 14217
(716) 875-9501

**Task Force Omega of North
Carolina**
c/o Bob and Sue Armstrong
7300 Chatterbird Ct.
Charlotte, NC 28226
(704) 542-5728
Fax: (704) 542-7319

**Task Force Omega of
Ohio**
c/o Tom McGraw
931 Bellflower SW
Canton, OH 44710
(216) 452-2285

**Task Force Omega of South
Carolina**
c/o A.R. "Snake" Fitzgerald
605 Level St.
Rock Hill, SC 29730-4744
(803) 328-5086
Fax: (803) 328-5086

Task Force Omega of Southern California
c/o John and Pat Pagel
803 Millburgh Ave.
Glendora, CA 91740
(626) 331-4010

Task Force Omega of Texas
c/o Becky Cordrey
1209 Rogers Pl.

Irving, TX 75060
(214) 986-1736

Task Force Omega of West Virginia
c/o Mary Alice and Jerry Bodeker
730 Daverson Rd.
Charleston, WV 25303
(304) 346-8475

Veterans of Foreign Wars of the United States (VFW)
406 W. 34th St.
Kansas City, MO 64111
(816) 756-3390
Fax: (816) 968-1157

Made up of American veterans of all overseas wars, the VFW works to ensure national security, assist the disabled as well as veterans' widows and dependents, and promote patriotism. Founded in 1899 and numbering more than 2.1 million members, it sponsors charitable and educational programs, maintains a museum, and advocates legislative action.

Publications: A monthly newsletter and *VFW Magazine*, which includes information on legislation, veterans' benefits, and news of local chapters.

Veterans of the Vietnam War
c/o Michael Milne
760 Jumper Rd.
Wilkes-Barre, PA 18702-8033
(717) 825-7215; (800) VIETNAM (toll-free)
Fax: (717) 825-8223
Web site: http://www.vvnw.org/vvnw/

This 20,000-member organization is made up of Vietnam veterans desiring to help and support one another. It runs a homeless veterans program, POW/MIA information and education program, speakers bureau, job training and apprenticeship program, and employment outreach program. It disseminates health information on Agent Orange and post-traumatic stress disorder. It also maintains a library, conducts research, and compiles statistics.

Publications: A quarterly magazine, *The Veteran Leader*, and a newsletter.

Veterans of the Vietnam War—State Organizations

Commander, State of Indiana
2321 Greentree N
Jeffersonville, IN 47129-8960

Commander, State of New York
1624 Chapin Ave.
Merrick, NY 11566-1944

Commander, State of Ohio
1267 U.S. Route 42
Ashland, OH 44805

Commander, State of Pennsylvania
43 Schumaher Ave.
Schuylkill Haven, PA 17972-2140

Veterans of the Vietnam War—Posts (U.S.)

California
Veterans of the Vietnam War, Inc. CA-06
1901 Talisman Dr.
Bakersfield, CA 93304

Delaware
Veterans of the Vietnam War, Inc. DE-01
P.O. Box 1302
Millsboro, DE 19966

Florida
Veterans of the Vietnam War, Inc. FL-04
4120 S.W. 31st Dr.
Hollywood, FL 33023

Georgia
Veterans of the Vietnam War, Inc. GA-01
P.O. Box 70355
Albany, GA 31707-0006

Veterans of the Vietnam War, Inc. GA-08
2031A Jasper Ct.
Ft. Gordon, GA 30905

Illinois
Veterans of the Vietnam War, Inc. IL-02
P.O. Box 48821
Niles, IL 60714-0821

Indiana
Veterans of the Vietnam War, Inc. IN-01
70 Vincennes St.
New Albany, IN 47151-0883

Veterans of the Vietnam War, Inc. IN-02
3828 E. Plymouth Rd.
Scottsburg, IN 47170

Veterans of the Vietnam War, Inc. IN-03
P.O. Box 7
Milltown, IN 47145

Veterans of the Vietnam War, Inc. IN-07
P.O. Box 391
Seymour, IN 47274

Veterans of the Vietnam War, Inc. IN-08
301 Main St.
Tell City, IN 47586

Veterans of the Vietnam War, Inc. IN-09
P.O. Box 181
Salem, IN 47167-0181

Veterans of the Vietnam War, Inc. IN-10
148 Wildwood Dr.
Madison, IN 47250

Kansas
Veterans of the Vietnam War, Inc. KS-01
P.O. Box 2701
Garden City, KS 67846-2935

Kentucky
Veterans of the Vietnam War, Inc. KY-02
3444 Cumberland Falls Hwy.
Corbin, KY 40701

Veterans of the Vietnam War, Inc. KY-03
185 B. Nelson Rd.
Melbourne, KY 41059

Veterans of the Vietnam War, Inc. KY-04
P.O. Box 817
Flatwoods, KY 41139

Veterans of the Vietnam War, Inc. KY-06
432 N. 25th St.
Louisville, KY 40212

Louisiana
Veterans of the Vietnam War, Inc. LA-04
P.O. Box 5913
Thibodaux, LA 70302

Michigan
Veterans of the Vietnam War, Inc. MI-03
2431 U.S. Hwy. 23S
Alpena, MI 49707-4615

Veterans of the Vietnam War, Inc. MI-05
P.O. Box 26
Allen Park, MI 48101

Veterans of the Vietnam War, Inc. MI-07
15775 Charles Rd.
Eastpointe, MI 48026-6577

Mississippi
Veterans of the Vietnam War, Inc. MS-01
813 E. Main St.
Tupelo, MS 38801

New Jersey
Veterans of the Vietnam War, Inc. NJ-03
P.O. Box 885
Jackson, NJ 08527-0885

New York
Veterans of the Vietnam War, Inc. NY-03
P.O. Box 7520
Freeport, NY 11520

Veterans of the Vietnam War, Inc. NY-10
22 Grove Place
Babylon, NY 11702

Veterans of the Vietnam War, Inc. NY-14
P.O. Box 521
Kings Park, NY 11754-0521

Ohio
Veterans of the Vietnam War, Inc. OH-01
P.O. Box 495T
Bridgeport, OH 43912

Veterans of the Vietnam War, Inc. OH-02
173 N. Wilson Rd., No. 10
Columbus, OH 43204-1226

Veterans of the Vietnam War, Inc. OH-03
P.O. Box 40803
Cincinnati, OH 45240

Veterans of the Vietnam War, Inc. OH-07
22971 Truman Ave.
Wickliffe, OH 44092

Veterans of the Vietnam War, Inc. OH-08
5648 Buenos Aires Blvd.
Westerville, OH 43081-4254

Veterans of the Vietnam War, Inc. OH-09
P.O. Box 1004
Ashland, OH 44805

Veterans of the Vietnam War, Inc. OH-18
P.O. Box 104
Jackson, OH 45640-0104

Pennsylvania
Veterans of the Vietnam War, Inc. PA-02
P.O. Box 2345
Wilkes-Barre, PA 18701-1052

Veterans of the Vietnam War, Inc. PA-22
255 Taylor Rd.
Julian, PA 16844

Veterans of the Vietnam War, Inc. PA-23
P.O. Box 114
Mt. Union, PA 17066

Veterans of the Vietnam War, Inc. PA-26
P.O. Box 921
Carlisle, PA 17013-5921

Veterans of the Vietnam War, Inc. PA-29
612 5th St.
Port Carbon, PA 17965-1102

Veterans of the Vietnam War, Inc. PA-30
P.O. Box 4292
Lancaster, PA 17604-2139

Veterans of the Vietnam War, Inc. PA-31
RR 1, Box 29
New Galilee, PA 16141

Veterans of the Vietnam War, Inc. PA-35
128–130 N. George St.
York, PA 17401

Veterans of the Vietnam War, Inc. PA-40
P.O. Box 182
Clarkes Summit, PA 18411-0182

Veterans of the Vietnam War, Inc. PA-41
1571 S. Main St.
Chambersburg, PA 17201

Veterans of the Vietnam War, Inc. PA-47
RR 2, Box 290A
Curwensville, PA 16833

Veterans of the Vietnam War, Inc. PA-51
P.O. Box 23
Hughesville, PA 17737-0023

Veterans of the Vietnam War, Inc. PA-52
637 Madison St.
Meadville, PA 16335

Veterans of the Vietnam War, Inc. PA-54
P.O. Box 218
Northampton, PA 18067-0218

Veterans of the Vietnam War, Inc. PA-56
706 E. Main St.
Nanticoke, PA 18634-1818

Veterans of the Vietnam War, Inc. PA-57
P.O. Box 694
Tobyhanna, PA 18466

Veterans of the Vietnam War, Inc. PA-59
1980 Chambersburg Rd.
Gettysburg, PA 17325

Veterans of the Vietnam War, Inc. PA-60
121 Center St.
Stockerton, PA 18083

Veterans of the Vietnam War, Inc. PA-61
P.O. Box 16953
Philadelphia, PA 19153-0953

Veterans of the Vietnam War, Inc. PA-65
305 Allendale Dr.
Morrisville, PA 19053-6215

Veterans of the Vietnam War, Inc. PA-66
c/o 500 Kennedy Blvd., No. 207
Pittston, PA 18640

Veterans of the Vietnam War, Inc. PA-73
51 N. Tulpehocken St.
Pine Grove, PA 17963-1214

Veterans of the Vietnam War, Inc. PA-75
P.O. Box 236
Macungie, PA 18062-0236

Veterans of the Vietnam War, Inc. PA-77
RR 3, Box 390
Bloomsburg, PA 17815

Virginia
Veterans of the Vietnam War, Inc. VA-08
P.O. Box 1524
Hopewell, VA 23860-1524

Washington
Veterans of the Vietnam War, Inc. WA-01
P.O. Box 27304
Seattle, WA 98125-1804

Veterans of the Vietnam War, Inc. WA-02
P.O. Box 5744
Bellevue, WA 98006-0244

Veterans of the Vietnam War, Inc. WA-03
P.O. Box 98409
Tacoma, WA 98498

Veterans of the Vietnam War, Inc. WA-04
P.O. Box 3126
Everett, WA 98203

Veterans of the Vietnam War, Inc. WA-05
P.O. Box 28764
Seattle, WA 98118-0764

Veterans of the Vietnam War, Inc. WA-06
2908 16th Ave. S
Seattle, WA 98144

Veterans of the Vietnam War, Inc. WA-10
P.O. Box 28922
Seattle, WA 98118-0922

Wisconsin
Veterans of the Vietnam War, Inc. WI-01
P.O. Box 14
New Glarus, WI 53574

Veterans of the Vietnam War, Inc. WI-02
2900 W. Lincoln Ave.
Milwaukee, WI 53215

West Virginia
Veterans of the Vietnam War, Inc. WV-03
P.O. Box 657
Bunker Hill, WV 25413

Veterans of the Vietnam War, Inc. WV-07
P.O. Box 917
Huntington, WV 25712-0917

Veterans of the Vietnam War—Posts (Canada)

Veterans of the Vietnam War, Inc. CN-01
775 Steeles Ave. W, No. 1605
Willowdale, Ontario
M2R 2S8 Canada

Vietnam Combat Veterans
1267 Alma Ct.
San Jose, CA 95112
(408) 288-6305
Web site: http://www.iinc.com/~flewhuey

Founded in 1980, this 375-member organization publicizes the plight of American servicemen who are POWs or MIAs in Southeast Asia and seeks their return. It sponsors a traveling display of the Vietnam Veterans Memorial Wall, conducts research on veterans' issues, offers seminars, and produces press releases on the POW/MIA issue. It also maintains an archive of relevant documents.

Vietnam Veterans of America Foundation
2001 S Street NW, Ste. 740
Washington, DC 20009
(202) 483-9222
Fax: (202) 483-9312
E-mail: vvaf@vi.org
Web site: http://www.vvaf.org

The Vietnam Veterans of America Foundation is "an international educational, humanitarian, and advocacy organization dedicated to addressing the causes, conduct, and consequences of war." Through its Institute for Conflict Resolution, it conducts symposia that teach the techniques of international dispute resolution to students and diplomats. Founded in 1980, it operates humanitarian programs in Southeast Asia, Africa, and Central America. In Southeast Asia it provides humanitarian assistance to the victims of war, operating the Kien Khleang Prosthetics and Rehabilitation Center in Kien Khleang, Cambodia, which manufactures and fits prosthetic limbs and wheelchairs. It also operates the Children's Orthotics Program at the Institute for the Protection of Children's Health in Hanoi, Vietnam. In 1991, the foundation launched an international and U.S. campaign to ban the production, sale, and use of antipersonnel land mines.

VietNow National
1835 Broadway
Rockford, IL 61104-5409
(815) 227-5100; (815) 227-5126; (800) 837-8669
Fax: (815) 227-5127
E-mail: Vnnatl@inwave.com

VietNow is made up of post-1957 U.S. veterans, who qualify for full membership, and interested civilians and pre-1957 veterans, who qualify for associate membership. VietNow aims to provide an organization through which veterans can help each other in such areas as health and employment. It maintains a speakers' bureau, through which it attempts to educate the public about the POW/MIA issue, works to pass laws to prevent the future abandonment of POWs, and helps individual families. The group was founded in 1980 and has 3,000 members.

Publication: A quarterly magazine, *VietNow,* which contains information on health, POW/MIA issues, and legislative efforts.

Selected Print Resources 6

General Reference Works

Dunn, Joe P. *Teaching the Vietnam War: Resources and Assessments.* Los Angeles: California State University, Center for the Study of Armament and Disarmament, Occasional Papers Series No. 18, 1990. 91pp.

A course outline and annotated bibliography written by a professor who teaches college courses on the Vietnam conflict. Chapter 3 contains entries about POW memoirs, descriptions of rescue attempts, polemics on the MIA issue, and novels on the subject.

Kutler, Stanley I., ed. *Encyclopedia of the Vietnam War.* New York: Charles Scribner's Sons, 1996. 711pp. ISBN 0-13-276932-8.

This is a comprehensive and up-to-date source of information on all aspects of the Vietnam war. All entries are supported by bibliographies.

Malo, Jean-Jacques, and Tony Williams. *Vietnam War Films.* London: McFarland, 1994. 567 pp. ISBN 0-89950-781-6.

This reference work provides production information, themes, and synopses for more than 600 nondocumentary films about the Vietnam war made between 1939 and 1992. Included are feature, made-for-television, pilot, and short movies. These films were made in the United States, Vietnam, France, Belgium, Australia, Hong Kong, South Africa, Great Britain, and elsewhere. They are indexed by chronology, country of origin, directors, writers, and selected actors.

Matray, James I., ed. *Historical Dictionary of the Korean War.* Westport, CT: Greenwood Press, 1991. 626 pp. ISBN 0-313-25924-0.

This dictionary includes entries on military forces, military operations and battles, and biographies of leaders. It includes as well a chronology of events, statistical information, and a selected bibliography. Each entry is followed by a list of suggested readings.

Olson, James S. *The Vietnam War: Handbook of the Literature and Research.* Westport, CT: Greenwood Publishing, 1993. 514 pp. ISBN 0-313-27422-3.

This comprehensive collection contains extensive bibliographic essays on general background and primary sources, military strategy, international relations, war crimes, peace negotiations, Indo-Chinese history, U.S. minority involvement, the antiwar movement, American literature, television, and film, and POWs and MIAs.

Peake, Louis A. *The United States in the Vietnam War: A Selected Annotated Bibliography.* New York: Garland, 1986. 406 pp. ISBN 0-8240-8946-4.

Includes some 1,500 citations on all aspects of Vietnam culture, history, and geography. As regards the war in Vietnam, it contains references to military actions, politics, the media, the domestic impact of the war, and the war's aftermath, as well as the war in art, music, and literature. The personal accounts of POWs and MIAs, as well as fictional works on that topic, also are included.

Sandler, Stanley, ed. *The Korean War: An Encyclopedia.* New York: Garland, 1995. 416 pp. ISBN 0-8240-4445-2.

This comprehensive reference work includes maps, a chronology of events, and a selected bibliography. Among others are entries on aircraft and armaments, major battles, biographies of leaders, historiography of the war, media coverage, and contributions of armed forces for different countries and of the branches of the U.S. military. Each entry is supported by a bibliography.

Schamel, Charles E., comp. *Records Relating to American Prisoners of War and Missing in Action from the Vietnam War Era, 1960–1994.* Reference Information Paper 90. Washington, D.C.: National Archives and Records Administration, 1996. 127 pp.

This work describes the contents and location of archival records of textual, electronic, audio, and videotaped material and motion pictures relating to POW/MIA questions. It includes records compiled by the Department of Defense, the Veterans Administration, and congressional committees, with live-sighting reports, action reports, casualty files, unit histories, depositions, and other testimony given to the latter.

Singleton, Carl. *Vietnam Studies: An Annotated Bibliography.* Lanham, MD: Scarecrow Press, 1997. 303 pp. ISBN 0-8108-3314-4.

This bibliography includes chapters on the history of Vietnam, its culture and arts, language and literature, and business and economics. Also treated are such topics as Vietnamese Americans, contemporary Vietnam, and the Vietnam conflict. Chapter 8 specifically addresses books on the POW/MIA issue.

Summers, Harry G., Jr. *Historical Atlas of the Vietnam War.* Boston: Houghton Mifflin, 1995. 224 pp. ISBN 0-395-72223-3.

Containing full-color maps and photographs of Vietnam war operations, this military description of the war is very readable. The author also provides historical background of U.S. participation in the war, details of particular battles, and a detailed chronology.

United States Air Force Academy. *A Revolutionary War: Korea and the Transformation of the Postwar World.* Special Bibliography Series No. 84. Washington, DC: U.S. Government Printing Office, 1992. 84 pp.

This selected, unannotated bibliography of Korean War materials includes books, government reports, journal articles, and technical reports. Among the topics covered are origins of the war, military forces, overview of ground, air, and naval operations, the Truman-McArthur controversy, POWs, peace negotiations, the armistice, and media coverage.

Wehrkamp, Tim, comp. *Records Relating to American Prisoners of War and Missing-in-Action Personnel from the Korean War and during the Cold War Era.* Reference Information Paper 102. Washington, DC: National Archives and Records Administration, 1997. 144 pp.

Provides descriptions of more than 190 series of textual, electronic, photographic, motion picture, sound recordings, and video recordings pertaining to servicemen listed as MIA from the Korean War and Cold War (1945–1991, excluding the Vietnam war, the records of which are handled separately) available in the National Archives. It includes records collected by the Department of State, U.S. Joint Chiefs of Staff, the Army Staff, the Office of the Secretary of Defense, the U.S. Air Force, and the Veterans Administration. Among the holdings are lists of repatriated prisoners of war, POW debriefings, and the Korean Conflict Casualty File.

Books

Alvarez, Everett, Jr., and Anthony S. Pitch. *Chained Eagle.* New York: D. I. Fine, 1989. 308 pp. ISBN 1-55611-167-3.

Navy pilot Everett Alvarez was captured on 5 August 1964 at Hon Gai, Vietnam. The longest held of the prisoners repatriated in 1973, he chronicles his experiences in Hanoi-area prisons, including being paraded through the streets of Hanoi and pummeled by a mob. He also briefly describes his homecoming and later life.

Anton, Frank, with Tommy Denton. *Why Didn't You Get Me Out? Betrayal in the Viet Cong Death Camps.* Arlington, TX: Summit Publishing Group, 1997. 196 pp. ISBN 1-56530-251-6.

Anton describes his three-year ordeal in a jungle POW camp in South Vietnam. After a forced march north on the Ho Chi Minh Trail, he endured another two years of internment in a Hanoi prison camp called the Plantation. In later chapters, he discusses his role in the court-martial of Robert Garwood for collaboration. He also gives his views on the POW/MIA issue, asserting that prisoners were left behind at war's end and that he and other POWs were not rescued because the United States did not want to jeopardize its spy network.

Blakey, Scott. *Prisoner of War: The Survival of Commander Richard A. Stratton.* Garden City, NY: Anchor Press, Doubleday, 1978. 397 pp. ISBN 0-38512-905-X.

On 5 January 1967, navy pilot Richard Stratton was taken prisoner after his plane crashed during a combat mission over North Vietnam. This is the story of his six-year captivity in Hanoi prisons. In addition, the work describes the stateside activities of Richard Stratton's wife, Alice, and others to secure the release of American POWs.

Brace, Ernest C. *A Code to Keep.* New York: St. Martin's Press, 1988. 264 pp. ISBN 0-31201-403-1.

A civilian pilot working for the Central Intelligence Agency, Ernie Brace was captured in May 1965 at a small airstrip in Laos. In this memoir, he recounts how he was held prisoner in a 3-by-4-foot bamboo cage for more than three years, how he made several unsuccessful escape attempts, and how he was moved to a prison camp in Hanoi. Although he was offered an early release, Brace declined it and became one of the main communication links in his prison. He was one of only nine prisoners captured in Laos who survived and returned to the United States.

Clarke, Douglas L. *The Missing Man: Politics and the MIA.* Washington, DC: National Defense University, 1979. 121 pp.

Clarke describes the process of determining the MIA status, the political and psychological considerations that impacted that process during the 1970s, and the development of the MIA question as an example of the intersection of domestic and international foreign policy. He analyzes the approaches to resolving the issue taken by Presidents Nixon, Ford, and Carter.

Coffee, Gerald. *Beyond Survival: Building on the Hard Times—A POW's Inspiring Story.* New York: G. P. Putnam's Sons, 1990. 287 pp. ISBN 0-399-13416-6.

This is one of many firsthand accounts of an American prisoner in Vietnam. Coffee describes his capture during the war and captivity in the Hanoi prison known as the Zoo. He writes of enduring torture and his forced participation in the infamous "Hanoi parade" during which he and other prisoners had to suffer a gauntlet of beatings by Vietnamese Communists. Finally he tells of his release in 1973 during Operation Homecoming.

Colvin, Rod. *First Heroes: The POWs Left Behind in Vietnam.* New York: Irvington Publishers, 1987. 365 pp. ISBN 0-8290-2008-X.

Writing from an activist's viewpoint, the author presents reasons why he believes prisoners were left behind at war's end, including testimony of a Vietnamese mortician who said he personally prepared hundreds of U.S. servicemen's remains, and the previous trend of Communists to keep prisoners of war for political purposes. Colvin describes the harsh treatment of prisoners and the opinions of Marine Pfc. Robert Garwood, who remained in Vietnam after 1973 and upon his return to the United States in 1979 was convicted of collaborating with the enemy; however, some people believe that Garwood was actually a prisoner. The first-person narratives of family members of MIAs make up the second half of the book.

Daly, James A., and Lee Bergman. *A Hero's Welcome: The Conscience of Sergeant James Daly versus the United States Army.* New York: Bobbs- Merrill, 1975. 267 pp. ISBN 0-67252-030-3.

This first-person account begins during Daly's high school years, when he objected to the Vietnam war on religious and moral grounds. Daly describes his efforts to receive conscientious objector status and undertake alternative service and how he was "tricked" by recruitment officials. When Daly entered military service and first saw combat, he was unable to fire on the enemy and was held captive, initially in a jungle camp in South Vietnam and later in North Vietnam in a Hanoi prison. After his release, he and others from the jungle camp were investigated on charges of misconduct for collaborating with the enemy, but the charges were eventually dropped.

Dengler, Dieter. *Escape from Laos.* San Rafael, CA: Presidio Press, 1979. 211 pp. ISBN 0-89141-076-7.

Dengler was one of only two men ever to escape from Laotian prison camps during the Vietnam war. In this account, he tells of the horrible conditions of the camp and the escape effort by all of the half-starved prisoners. Had it not been for a U.S. pilot's spotting of Dengler from the air, the wandering and weak Dengler likely would have died in the jungle.

Denton, Jeremiah. *When Hell Was in Session.* New York: Readers Digest Press, 1976; reprint, Mobile, AL: Traditional Press, 1982. 246 pp. ISBN 0-88349-112-5.

One of the most famous returning POWs tells his story of captivity, torture, and resistance. Denton was held in the Alcatraz prison camp in Hanoi and was forced to endure the infamous "Hanoi parade" during which he and other prisoners were forced through a gauntlet of Vietnamese who beat them.

Franklin, H. Bruce. *M.I.A. or Mythmaking in America.* New York: Lawrence Hill Books, 1992. 225 pp. ISBN 1-55652-118-9.

Working from the assumption that no live prisoners of war exist in Southeast Asia, Franklin treats the terminology used to discuss MIAs, the issue of deserters, the development of POW/MIA as an issue, and the positions and activities of the Ford, Carter, and Reagan administrations. Franklin asserts that the issue has taken on a mythic function in American culture, and he attempts to disprove the assertions of the POW/MIA activists.

Howes, Craig. *Voices of the Vietnam POWs: Witnesses to Their Fight.* New York: Oxford University Press, 1993. 295 pp. ISBN 0-19507-630-3.

Howes evaluates how different interpretations of the Code of Conduct created controversy among prisoners during the Vietnam war. He describes how the senior ranking officers in charge of the prisoners in Hanoi prison camps lived and how their decisions affected the group of POWs as a whole. Finally, Howes discusses the effect of religious beliefs on the prisoners and how returned prisoners recounted their stories in different ways. He tells the personal stories of Richard Stratton, Doug Hegdahl, Robert Garwood, James Stockdale, and the prisoners of the Communist jungle camps in South Vietnam.

Jensen-Stevenson, Monika, and William Stevenson. *Kiss the Boys Goodbye: How the United States Betrayed Its Own POWs in Vietnam.* Toronto: McClelland and Stewart, 1990. 493 pp. ISBN 0-77108-326-2.

Monika Jensen-Stevenson was a producer of the CBS news magazine *60 Minutes* in the mid-1980s when she began to investigate allegations that American POWs had been left in Southeast Asia following the release of prisoners in 1973. She describes her research efforts, interviews with many people, and the warnings she received from U.S. government officials to drop her inquiries. Those interviewed include U.S. Marine Private Robert Garwood, who was convicted of collaboration after he returned to the United States in 1979, but who some people believe to have really been a prisoner of the Vietnamese. The authors disclose the opinions of former National Security Agency analyst Jerry Mooney, who had tracked the movements of U.S. aircrewmen after their capture by the Communists and knew of POWs who had been held after 1973 and others who were executed. Mooney maintained that prisoners with particularly important technical knowledge were sent to Russia. Jensen-Stevenson also recounts her dealings with Texas millionaire Ross Perot, who was given a security clearance by President Reagan and read many classified documents on the MIA issue, only to have his clearance rescinded by President Bush after Perot accused the U.S. government of using its sections of the agency for POW/MIA affairs as a cover for covert military operations in Southeast Asia funded by drug smuggling. According to the authors, Scott Barnes, who they describe as a "freelance specialist in covert warfare," declared that while on a mission in Southeast Asia for the Drug Enforcement Agency he saw American prisoners and was ordered to kill them. Other interviewees include former U.S. Navy Captain Eugene "Red" McDaniel, who operated the American Defense Institute and accused the National League of Families of American Prisoners and Missing in Southeast Asia of discouraging questions about administration MIA policy. About private efforts to rescue POWs, the authors tell of the ill-fated missions of former Green Beret Bo Gritz during the mid-1980s and of Major Mark "Zippo" Smith and Sergeant Melvin McIntire, who were involved in a supposed rescue attempt of American POWs. In conclusion, Jensen-Stevenson asserts that the U.S. government possesses detailed information about POWs left in Southeast Asia but chose to keep this information secret in order to maintain its intelligence network.

Johnson, Sam, and Jan Winebrenner. *Captive Warriors: A Vietnam POW's Story.* College Station: Texas A&M University Press, 1992. 310 pp. ISBN 0-89096-496-3.

This is one of the most recent of many POW memoirs. Johnson describes his years at the Hanoi prison camp known as Alcatraz.

Keating, Susan Katz. *Prisoners of Hope: Exploiting the POW/ MIA Myth in America.* New York: Random House, 1994. 276 pp. ISBN 0-679-43016-4.

A reporter for the *Washington Post* covering the MIA story, Keating at first believed in the existence of live prisoners of war in Southeast Asia. However, after years of investigating, she came to believe that because the U.S. government had mishandled this important issue, the myth of live prisoners of war had sprung up. Keating discusses the statistics, the activities of the Defense Intelligence Agency (DIA), deserters and collaborators, the possibility of a cover-up by the U.S. government, and the questionable forensic procedures used by the Central Identification Laboratory in Hawaii. The "true conspiracies," according to Keating, are efforts by MIA activists, con artists, information peddlers, and mercenaries who employ the issue for personal gain, monetary or otherwise.

McConnell, Malcolm, with research by Theodore G. Schweitzer III. *Inside Hanoi's Secret Archives: Solving the MIA Mystery.* New York: Simon and Schuster, 1995. 462 pp. ISBN 0-671-87118-8.

McConnell describes the covert operation known as Operation Swamp Ranger through which Vietnamese archivists at the Central Military Museum of the People's Army of Vietnam (PAVN) in Hanoi funneled important documents to the United States through Theodore Schweitzer, a private contractor supposedly researching the issue to write a book. Among these documents was a copy of the Red Book (a listing of archive contents), photographs of remains of U.S. military personnel, and dog tags and identity cards of MIAs. Although many of these items were later found to be useful in partially resolving discrepancy cases, the Vietnamese did hold back certain items, such as individual prisoner records that included audiotapes of interrogations. Schweitzer recounted anecdotal evidence to suggest that these withheld documents might relate to prisoners who were tortured to death, executed, or dismembered by mobs. McConnell puts this operation in context by describing the evolution of the POW/MIA issue—including conspiracy theories—and the activities of the Joint Task Force–Full Accounting (JTF-FA).

McCubbin, Hamilton I., Barbara B. Dahl, Philip J. Metres Jr., Edna J. Hunter, and John A. Plag, eds. *Family Separation and Reunion: Families of Prisoners of War and Servicemen Missing in Action.* San Diego, CA: Center for Prisoner of War Studies and Washington, DC: U.S. Government Printing Office, 1974. 234 pp.

The scientific studies in this collection draw from a variety of disciplines. Investigators studied such topics as the emotional, social, and legal problems of families of POWs and MIAs. Later chapters deal with the medical problems, adjustment, and reintegration into society of repatriated prisoners and the medical, counseling, and legal services provided to the families of former prisoners and MIAs.

Norman, Geoffrey. *Bouncing Back.* Boston: Houghton Mifflin, 1990. 248 pp. ISBN 0-395-45186-8.

Norman focuses on navy pilot Al Stafford, who was shot down in 1967 and held for five years in various Hanoi-area prisons. This work tells of members of the Fourth Allied POW Wing, who resisted their captors through a strategy that they called "bouncing back." They include pilot Richard Stratton, who was the senior ranking officer at the Plantation, and seaman Douglas Hegdahl. Hegdahl had purposefully convinced the Vietnamese that he was a "stupid peasant," though he had a phenomenal memory. He memorized the names of all of the prisoners in his camp (some 260). He was ordered by his commanding officer to accept early release in order to disclose the prisoners' names and tell the U.S. government of the poor conditions and barbarous treatment of American prisoners by the Vietnamese Communists. The standing order was for all prisoners to deny early release and be repatriated as a group.

O'Daniel, Larry J. *Missing in Action: Trail of Deceit.* New Rochelle, NY: Arlington House, 1979. 304 pp. ISBN 0-87000-450-6.

This early activist account focuses on the activity of the Carter administration, including the Woodcock Commission, which concluded that no live prisoners remained in Southeast Asia as a result of the war in Vietnam. O'Daniel charges the U.S. government with hiding evidence that live POWs were being held in Southeast Asia. He also criticizes President Carter's attempts to normalize relations with the Socialist Republic of Vietnam.

Risner, Robinson. *The Passing of the Night: My Seven Years as a Prisoner of the North Vietnamese.* New York: Random House, 1973. 264 pp. ISBN 0-394-48967-5.

His aircraft shot down in 1965, Colonel Robinson Risner was held in the Hanoi Hilton until his release during Operation Homecoming in 1973. This is his first-person account of torture and indoctrination as well as resistance by U.S. prisoners.

Rowan, Stephen A. *They Wouldn't Let Us Die: The Prisoners of War Tell Their Story.* Middle Village, NY: Jonathan David Publishers, 1973. 252 pp. ISBN 0-82460-157-2.

Rowan provides an overview of the conditions and treatment American prisoners suffered. In addition, he includes interviews he conducted with a dozen former prisoners shortly after their repatriation in 1973. In these interviews, the former prisoners discuss the Code of Conduct, how they survived torture, and how the prison communication system worked and why this communication system was so important to maintaining morale.

Sauter, Mark, and Jim Sanders. *The Men We Left Behind: Henry Kissinger, the Politics of Deceit and the Tragic Fate of POWs after the Vietnam War.* Washington, DC: National Press Books, 1993. 320 pp. ISBN 1-882605-03-9.

Declaring that Americans were knowingly left in Southeast Asia after 1973, the authors assert that some so-called deserters were actually prisoners and cite previous experiences of French forces during the Indochina War (1946–1954) and Americans during the Korean War. They also cite instances in which they believe an American was taken prisoner but declared dead by U.S. officials. They describe efforts made during the war to swap prisoners, setting a precedent for the North Vietnamese to expect the United States to pay ransom. The authors describe negotiations with the North Vietnamese that led to the Paris Peace Accords and maintain these efforts centered on payment of ransom in the form of aid to Vietnam for POWs. They suggest that Richard Nixon and Henry Kissinger later prevented tape recordings and documents about POW policy and negotiations from becoming public to hide the fact that they were not willing to ransom the prisoners kept back after Operation Homecoming. They also contend that the U.S. government ignored and covered up evidence of prisoners held after 1973 for political reasons. They describe the activities of the DIA and the possibility that some POWs were sent to the Soviet Union. They criticize the congressional committees that investigated the issue, including the Montgomery Committee of the 1970s and the Senate Select Committee on POW/MIAs. Sanders and Sauter, along with R. Cork Kirkwood, cover much of the same material in *Soldiers of Misfortune: Washington's Secret Betrayal of America's POWs in the Soviet Union* (Washington, DC: National Press Books, 1992).

Stern, Lewis M. *Imprisoned or Missing in Vietnam: Policies of the Vietnamese Government Concerning Captured and Unaccounted for United States Soldiers, 1969–1994.* Jefferson, NC: McFarland, 1995. 191 pp. ISBN 0-7864-0121-4.

Based on American and Vietnamese government and public documents, this work describes in detail the activities of the U.S. and Vietnamese agencies tasked with accounting for American MIAs, including the many technical meetings and investigations that took place. Stern discusses the negotiating behavior and positions of the Vietnamese on the POW/MIA issue from the early postwar years to 1994, when the trade embargo against Vietnam was lifted by President Clinton. Vietnamese tactics include the many public statements made by the Vietnamese to support their position of cooperation and their moral superiority in demanding that the United States honor its commitment to "healing the wounds of war" in Vietnam through humanitarian assistance.

Stockdale, James B. *A Vietnam Experience: Ten Years of Reflection.* Stanford, CA: Hoover Institution, 1984. 147 pp. ISBN 0-81798-151-9.

This is a chronologically arranged collection of articles, speeches, and essays Stockdale, a prisoner of the North Vietnamese for eight years, made from 1973 to 1983. Stockdale treats such topics as his POW experiences, leadership, freedom, and duty.

Veith, George J. *Code-Name Bright Light: The Untold Story of U.S. POW Rescue Efforts during the Vietnam War.* New York: Free Press, 1998. 408 pp. ISBN 0-68483-514-2.

Based on numerous declassified intelligence documents and interviews with military personnel involved in the efforts of the Joint Personnel Recovery Center (JPRC), Veith reveals the efforts of the U.S. government's intelligence programs to locate and rescue prisoners of war. These included the efforts of the Military Assistance Command, Vietnam's Studies and Observation Group (MACV-SOG), the JPRC, the State Department, the DIA, and the CIA. Bright Light was the unclassified code name given to information related to American or friendly military or civilian prisoners. Despite the efforts of these many groups, no American prisoner of war was successfully rescued during the Vietnam war.

Zalin, Grant. *Survivors.* New York: W. W. Norton, 1975. 345 pp. ISBN 0-39308-727-1.

Zalin recounts the capture and imprisonment in jungle POW camps and later Hanoi prisons of pilot Frank Anton; riflemen David Harker, Jim Strickland, and James Daly; grenadier Willie

Watkins; mortarmen Isaiah McMillan and Tom David; special forces operative John Young; and others who did not survive captivity. Several of these soldiers were among those accused of collaborating with the North Vietnamese by pilot Theodore Guy, who also tells his story here. Eventually the military dropped the charges of collaboration.

Government Reports

U.S. House of Representatives. *Accounting for U.S. POW/MIAs in Southeast Asia.* Hearing before the Military Personnel Subcommittee, Committee on National Security, 104th Congress, first session, 29 June 1995. Washington, DC: U.S. Government Printing Office, 1996. 866 pp.

Contains statements by family members of POWs; retired colonel of the Air Force Ted Guy (a former POW); the directors of the National Alliance of Families, the American Coalition Legion, and the National Vietnam Veterans; the commander of the Joint Task Force–Full Accounting, and the assistant secretary of state for East Asia and Pacific Affairs, among others. These witnesses describe government accounting activities, list priority discrepancy cases, present arguments for and against normalization of relations with Vietnam, and deal with individual MIA cases.

U.S. House of Representatives. *Department of Defense's Comprehensive Review of POW/MIA Cases.* Hearings before the Military Personnel Subcommittee, Committee on National Security, 104th Congress, first session, 20, 30 November 1995. Washington, DC: U.S. Government Printing Office, 1996. 81 pp.

James W. Wold, then deputy secretary of defense (POW/MIA Affairs), describes the motivation, methodology, results, and next steps to be taken as a result of the 1995 review of POW/MIA cases in Vietnam, Cambodia, and Laos. Wold conducted the case-by-case review after becoming head of DPMO in 1993. He summarized by case what had been done to date on a file and what the Vietnamese government had provided on a case (citing recent document turnovers); predicted whether or not Vietnamese, Cambodian, or Laotian governments were likely to possess more information on a case; and decided whether to continue to actively investigate a case or defer it pending the acquisition of further information.

U.S. House of Representatives. *Examination of Operations at DIA's Special Office for Prisoners of War and Missing in Action.* Hearing before the Subcommittee on Asian and Pacific Affairs of the Committee on Foreign Affairs, 102d Congress, first session, 30 May 1991. Washington, DC: U.S. Government Printing Office, 1991. 54 pp.

Investigates complaints about the activities of the DIA POW/MIA Office, including the dramatic resignation of the DIA POW/MIA Office head, Colonel Millard Peck. In 1993, this office was replaced by the Department of Defense Prisoner of War/Missing in Action Office (DPMO).

U.S. House of Representatives. *POW/MIA: Where Do We Go from Here?* Hearing before the Subcommittee on Asian and Pacific Affairs of the Committee on Foreign Affairs, 103d Congress, second session, 10 February 1994. Washington, DC: U.S. Government Printing Office, 1994. 365 pp.

Representatives of veterans organizations and the National League of Families criticize President Clinton's lifting of the trade embargo of Vietnam and of the move toward normalization in general. U.S. government representatives, including Ambassador Winston Lord and JTF-FA Commander Thomas Needham, describe recent U.S. efforts to account for MIAs. Several witnesses debate the authenticity and content of the Quang "1205" document, which seemed to suggest that hundreds of American prisoners of war were retained in Southeast Asia after Operation Homecoming; the possibility of American POWs existing to this day; and the quality of work done by JTF-FA.

U.S. House of Representatives. *POWs/MIAs: Missing Pieces of the Puzzle.* Hearing before the Subcommittee on Asian and Pacific Affairs of the Committee on Foreign Affairs, 103d Congress, first session, 14 and 22 July 1993. Washington, DC: U.S. Government Printing Office, 1994. 259 pp.

Professor Stephen J. Morris describes the discovery in Russian archives and the contents of what became known as the Quang "1205" document (noted above). Jim Sanders (author with Jim Sauter of *The Men We Left Behind*) gives his opinions on the theory that POWs were withheld and sent to the Soviet Union. Others debate the veracity of the Quang "1205" document and its contents.

U.S. House of Representatives. *The Toon Mission to Russia: New Information of POW/MIAs.* Hearing before the Subcommittee on Asian and Pacific Affairs of the Committee on Foreign Affairs, 102d Congress, second session, 1 July 1992. Washington, DC: U.S. Government Printing Office, 1993. 83 pp.

This hearing deals with the efforts of the U.S.-Russian Joint Commission on POW/MIAs, headed by former ambassador to the Soviet Union Malcolm Toon and his Russian counterpart General Volkogonov. Toon stated that despite the support of Russian President Boris Yeltsin, who had previously indicated that American POWs had been held in Russia, he found "no evidence that could confirm that American POWs are still held in Russian territory." Director of the National League of Families of American Prisoners and Missing in Southeast Asia Ann Mills Griffiths provides a chronology of POW/MIA agreements between the United States and the Socialist Republic of Vietnam.

U.S. House of Representatives. *United States and Vietnamese Government Knowledge and Accountability for U.S. POW/MIAs.* Hearing before the Military Personnel Subcommittee, Committee on National Security, 104th Congress, first session, 14 November 1995. Washington, DC: U.S. Government Printing Office, 1996. 266 pp.

Contains the statements of family members of POWs, the former director of Asian Affairs, National Security Council member Richard Childress, National League of Families of American Prisoners and Missing in Southeast Asia Director Ann Mills Griffiths, and Congressman Robert Dornan. Many of these witnesses testify that the government of Vietnam continues to withhold remains and documents, such as individual prisoner files, that would account for MIAs on the discrepancy case—that is, those who were known to be in the Vietnamese prison system at some time during the war.

U.S. Senate. *Oversight Hearings: Department of Defense, POW/MIA Family Issues, and Private Sector Issues.* Hearings before the Select Committee on POW/MIA Affairs, 102d Congress, second session, 1–4 December 1992. Washington, DC: U.S. Government Printing Office, 1994. 1,674 pp.

Investigates the effectiveness of the DIA POW/MIA Office. The hearings delve into the activities of veterans and activist organizations and review individual MIA case files.

U.S. Senate. *POW/MIA's: Report of the Select Committee on POW/MIAs.* Report 103-1, 103rd Congress, first session. Washington, DC: U.S. Government Printing Office, 1993. 1,223 pp.

The result of a yearlong congressional investigation (1991–1992), this report describes the purpose, approach, and methods of the Select Committee on POW/MIAs and summarizes its findings and recommendations. The full hearings of the committee were published separately in several volumes. The committee investigated issues related to the Paris Peace Accords, the accounting process, the Defense Intelligence Agency, and cooperation from the governments of Southeast Asia. It reviewed as well private activities on the issue and information from Russia, Korea, and China dating from the Korean War and the Cold War.

Periodical

The US Veteran Dispatch. Published by Ted Sampley (P.O. Box 246, Kinston, NC 28502), this periodical presents the views of activists on a wide variety of POW/MIA-related issues. It is published on the Web at http://www.usvetdsp.com.

Selected Nonprint Resources 7

Following is a sampling of the types of nonprint material that exist on the POW/MIA issue. By far the most dynamic source of information is the Internet, where information can be, and often is, altered frequently. Among the sites most useful to researchers are those that contain archives of documents that are often available only at document depository libraries.

Audio Book

Coffee, Gerald. *Beyond Survival: Building on the Hard Times—A POW's Inspiring Story.* New York: G. P. Putnam's Sons, 1990. ISBN 1-55525-332-6.

This is the audio version of Coffee's 1990 book (see Chapter 6), a firsthand account of his captivity during the Vietnam conflict.

Computer Programs

Operation Smoking Gun
Type: IBM-compatible only
Medium: 3.5- or 5.25-inch high-density disk (requires 4 megabytes of storage space on the hard drive)
Price: $40

Source: P.O.W. Network
              Box 68
              Skidmore, MO 64487-0068
              (660) 928-3304

This program contains the full documents, not summaries, of reported "Bright Light" sightings of American servicemen that were received by the U.S. government. Bright Light was the unclassified term referring to live American prisoners of war. The program also contains oral histories and interviews with former enemy soldiers, and wartime enemy documents that have been captured and translated. Included are documents written by former National Security Agency analyst Jerry Mooney and "POW and Politics," a paper written by George J. Veith and Bill Bell. All documents have been converted into ASCII and are indexed for easy retrieval. Alternatively, this information can be downloaded from the P.O.W. Network home page at http://222.asde.com/~pownet/

P.O.W. Biography Database
Type: IBM-compatible only
Medium: 3.5- or 5.25-inch high-density disk (requires 13 megabytes of storage space on the hard drive)
Price: $45
Source: P.O.W. Network
            Box 68
            Skidmore, MO 64487-0068
            (660) 928-3304

This program contains information from classified U.S. government documents. Each record has been compiled from more than a dozen sources of information on each of the officially listed 3,632 prisoners of war and missing in action servicemen from the Vietnam conflict. Minimally there is for each record: last name, first name, branch of service, rank at time of loss, race, date of loss, home state of record, DIA category, status, vehicle (such as aircraft piloted), and country of loss. Date of birth and capture coordinates are also included for some records. Of those 3,632 records, more than 2,820 have additional incident/biographical information. Each is noted for status (returnee, body recovered, still missing, AWOL, etc.). The database is searchable (for example, by filtering you can find all Air Force personnel listed in an F105D in North Vietnam from New Jersey, if you want). Screen search capability is excellent, and those records with biographies can be viewed on screen or printed. Each returnee has a release date noted, and each remains returned has an identification date. In addition, whereas many of the POW/MIA biographies are strongly military and incident related, returnee biographies include more personal information, such as information on families, rank at retirement or separation, and reflections and stories by the former prisoners. A few biographies are only a paragraph; others run to ten pages, but most are one or two pages long.

The Wall
Type:　　IBM-compatible only
Medium:　3.5- or 5.25-inch high-density disk (requires 20 megabytes of storage space on the hard drive)
Price:　　$55
Source:　P.O.W. Network
　　　　　Box 68
　　　　　Skidmore, MO 64487-0068
　　　　　(660) 928-3304

The Wall program contains information from U.S. government declassified documents as they relate to casualties from the Vietnam conflict that are listed on the Vietnam Veterans Memorial Wall in Washington, D.C. Minimally each record contains last name, first name, date of casualty, date of birth, hometown of record, home state, line and panel on the Wall, and branch of service. The program allows searches by last name, date of casualty, or hometown. Casualties by state can be printed with a keystroke. Print screen functions allow printing of data. No biographical information is available on any record.

Internet Sites

It is important to note that some Internet addresses are case sensitive, which means that the address must be input exactly as it is printed here, with capital and lowercase letters, slashes, periods, or other symbols.

**Advocacy and Intelligence Index for
Prisoners of War–Missing in Action**
http://www.aiipowmia.com

This site contains an archives and issues section, a news release section that features daily and weekly updates on issues, an archive of previous news releases, an updates section of events, and an E-mail server to send subscribers daily news and notices by E-mail. The archive contains such documents as President Nixon's secret letter to the Vietnamese, a transcript of Robert Garwood's court-martial for collaborating with the enemy, a chronology of the Vietnam war and POW/MIA issue from 1965 to 1992, a bibliography of several hundred POW/MIA-relevant books, definitions of military status, the Geneva Convention (1949) on the treatment of prisoners of war, and the joint resolution of Congress after the Gulf of Tonkin incident. Documents relating to the operations of the DIA POW/MIA Office include the Gaines Report, the Brooks Memorandum, the Tighe Task Force report, Col. Millard Peck's resignation letter from the POW/MIA Office, and the comprehensive review of the DIA. About the Korean conflict, the archive contains Department of Defense briefings, individual incident reports, and the Rand Corporation report on POW/MIA issues, which dealt with the possibility that Korean War POWs had been transferred to the Soviet Union and Soviet satellites. Other documents addressing the POW/MIA issue include the 1996 report of the U.S.-Russia Joint Commission and the Quang "1205" document. Congressional testimony includes that of "The Mortician" in June 1980, the Senate Select Committee Hearings from 1991 and 1992, and hearings before the House Veterans Affairs Committee in 1991 and before the House Subcommittee on Military Personnel in 1995 and 1996. A description of POW/MIA-related legislation in the House of Representatives and Senate during the 104th and 105th Congresses also is included.

Canadian POW/MIA Information Centre
http://www.ipsystems.com/powmia/

This site contains an archive of the center's newsletters, a listing of the names on and a photo album about the Vietnam Veterans Memorial in Washington, D.C., and excerpts from such works as *Soldiers of Misfortune* (Sanders, Sauter, and Kirkwood 1992) and *The Men We Left Behind* (Sauter and Sanders 1993). Other articles discuss the shootdown of the Baron 52 aircraft, the Lima Site 85 radar installation, and the possibility of POWs being sent to the Soviet Union. Former POW Theodore Guy describes his experiences in captivity, and the cases of individual MIAs, such as Charles Shelton and Gene DeBruin, are discussed. Critics allege the incompetence of the Department of Defense, particularly the Central Identification Laboratory in Hawaii, and decry the U.S.

government's treatment of Laotians who aided U.S. forces during the war. An article presents the facts about known deserter Earl Clyde Weatherman, while in another former major Mark Smith defends Robert Garwood, who was convicted in U.S. military court of collaborating with the enemy.

Defense Prisoner of War/Missing in Action Office (DPMO)
http://www.dtic.mil/dpmo/

This government Web site provides weekly updates that include information on accounting for the remains of MIAs, such as statistics, updated on a monthly basis, on the latest cases solved and schedules for family briefings. It publishes JTF-FA updates, special reports of activities of joint commissions on POW/MIA affairs, and debriefings of former POWs. It lists a schedule of POW/MIA-related events, such as technical talks and excavations in Vietnam, Laos, and Cambodia.

Freedom Flight Tribute Page
http://www.wardogs.com/fflight/html

This site contains a list of prisoners of war who are deceased. It also includes a "hall of shame" that lists phony prisoners of war (as of 2 April 1998, this list contained approximately 100 names). In addition, the site contains a bibliography of books by and about former prisoners of war and on the POW/MIA issue, and escape stories of several prisoners.

Hawk's Page
http://www.lakeozarks.net/~tedguy/

This is Theodore Guy's home page. Guy is a former prisoner of war who until the early 1990s did not believe that POWs still exist in Southeast Asia. However, after researching the topic, he changed his opinion. Here Guy presents the reasons why he believes POW could still be alive, including summaries of MIA cases accompanied by his opinions. Also included is an article by Robe P. Thompson regarding a particular MIA case.

Historical Text Archive: Vietnam War
http://sunsite.unc.edu/pub/academic/history/marshall/military/vietnam/

This extensive archive of military history dealing with many wars in which the United States played a role contains much useful information on the Vietnam war, including a short military history of the war and an archive of wartime photographs that includes photos of POWs. It contains the combat casualty files, which include such information as name, rank, date of casualty, home of record, location on the Vietnam Veterans Memorial Wall, and a few other data fields for the 58,000 casualties. A Peace Agreement subheading gives transcriptions of original documents, including Eisenhower's and Kennedy's letters to South Vietnamese president Ngo Dinh Diem, President Johnson's message to Congress after the Tonkin Gulf incident, the Paris Peace Accords, press releases made by Henry Kissinger and President Nixon, and John Kerry's statement to Congress in 1971 as head of the Vietnam Veterans Against the War. A bibliography of books relating to the U.S. Navy during Vietnam contains a section on POWs. The site also contains a glossary of military terms used during the war. The POW-MIA subheading contains the text (in condensed or expanded format) of the entire hearings of the U.S. Senate Select Committee on POW/MIA Affairs (known as the Senate Select Committee cochaired by John Kerry and Robert Smith) from 1991 and 1992. This site also offers reprints from *Vietnam Generation*, a quarterly journal/newsletter, and *Nam Vet*, a monthly newsletter dedicated to Vietnam veterans that includes articles on POW/MIA affairs. The archive also contains a military history of the Korean War and a collection of accounts by those serving in small combat units.

Korean War MIA/POW Help Desk
http://www.kimsoft.com/korea/mia-kr.htm
This site offers reprints of articles about Korean War POWs in the popular press, such as the *New York Times, Readers Digest, New American, U.S. News and World Report,* and the *Korean Herald.* It also includes a U.S. Department of Defense briefing about a POW/MIA fact-finding mission to Pyonyang, North Korea, in June 1996. The site contains links to related sites.

Korean War Project
http://www.koreanwar.org
Includes articles and recollections by Korean War veterans as well as a KIA/MIA database of the 33,642 servicemen officially declared dead by hostile means. It contains the U.S. government Korean War Casualty File from 1950 to 1953, which lists the names of those killed in action or missing. It offers as well photographs of the Korean War Veterans Memorial in Washington, D.C., and provides a bulletin board service.

National Alliance of Families Bits and Pieces Newsletter
http://www.nationalalliance.org/

This is the home page of the National Alliance of Families for the Return of America's Missing Servicemen, an activist organization. It presents information on legislative actions, provides the organization's positions on issues, encourages action by interested citizens, and responds to government press releases.

PBS/P.O.V. Re: Vietnam Stories since the War
http://www.pbs.org/pov/stories/

This is an oral history project that solicits stories from readers. The site was inspired by a Public Broadcasting Service (PBS) broadcast about Maya Lin, the designer of the Vietnam Veterans Memorial Wall in Washington, D.C. The featured personal stories selections change regularly. An archive keeps these stories, allowing them to be retrieved. The archive is searchable by concept or keyword.

POW/MIA Database (Correlated and Uncorrelated Information Relating to Missing Americans in Southeast Asia)
http://lcweb2.loc.gov/powhome.html

This bibliography of some 200,000 records from the Library of Congress allows researchers to use a variety of search methods (last name, country name, service branch, keywords) to locate documents that they can then order from the Library of Congress on microfilm or as photocopies. Documents contained in the database deal with treatment, location, and condition information on some 1,680 missing personnel from the Vietnam war. Holdings include intelligence reports, service records, loss incident reports, loss investigations, refugee reports, past and current recovery efforts, case analysis, loss site excavation reports, and archival documents found in Vietnam, Cambodia, and Laos. Although much information deals with specific cases, other documents deal with POW/MIA policy, release negotiations, diplomatic activities, and congressional hearings. The database, which includes information from the Department of Defense, the Joint Task Force–Full Accounting, the Department of Defense Prisoner of War/Missing in Action Office, the Central Identification Laboratory in Hawaii, the Pacific Command, and other sources, was supplied as a result of the McCain Act and Executive Order 12812. This database does not include information on POWs and MIAs from World War II, the Korean War, or the Cold War; POWs who were repatriated prior to or as part of Operation Homecoming; POWs who did not survive their imprisonment and whose remains were returned to the United States before December 1991; and unaccounted-for personnel after December 1991 whose records were withheld under the provision of the McCain Act that allowed families to hold private such information.

P.O.W. Network
http://www.asde.com/~pownet

Under the heading Operation Smoking Gun, this site contains 1,500 pages of declassified Bright Light live-sighting reports dated during the Vietnam war. (As noted earlier, Bright Light was the unclassified code name for captured U.S. military personnel.) It also includes translated enemy documents, oral histories, and interviews with former North Vietnamese soldiers, a transcript of the German-made documentary film about American POWs *Pilots in Pajamas,* and biographies of POWs and MIAs.

POW/MIA Forum
http://pages.prodigy.com/powforum/pwmiaf.htm

This site acts as a server for Web sites created by individuals in what is the Operation Just Cause Web ring. It links to some 300 sites, many of which are tributes to missing servicemen

S.E.A.R.C.H. Inc.
http://www.iwc.com/POW-Search/

This is the home page for the S.E.A.R.C.H. Inc. organization. It includes such documents as the Quang "1205" document, "Background Paper: Accountability of Missing Americans from the Korean War, Live Sighting Reports," "Top Secret POW Program" (about World War II POWs), a glossary of Vietnam veterans' terminology and slang, and the texts of several recent live-sighting reports from Southeast Asia that were received by the DPMO. This home page also lists links to over 400 other MIA-related sites, many of them providing tributes and information on individual MIAs.

U.S. POW/MIAs in Southeast Asia
http://www.wtvi.com/wesley/powmia/powmia.html

This site includes book reviews of POW/MIA activists' books, including an interview with Carol Hrdlicka, wife of David

Hrdlicka, MIA, in which she harshly criticizes the DIA and the U.S. government, and the book *Prisoners of Hope* by Susan Katz Keating (see Chapter 6). Included is a 28 April 1997 memo by Dr. Castle, chief of Southeast Asia archival research for the DPMO. Castle criticizes the archival research methods and interviews conducted by certain JFT-FA employees regarding the loss of LS-85, a radar site in Laos that accounted for 11 MIAs. Castle makes recommendations to improve the research process.

Vietnam Online
http://www.pbs.org/wgbh/pages/amex/vietnam/index.html

This is a companion site to the *Vietnam War on Videos* series created by the Public Broadcasting System. It includes a concise overview of the MIA issue. It also offers a detailed chronology of the era, short biographies of important persons during the war, and a dozen reflective personal essays (two by women). A reference section supplies maps of the war during various periods, a list of acronyms unique to the Vietnam conflict, and a listing of primary government documents on U.S. government policy on Vietnam.

Vietnam Veterans Home Page
http://grunt.space.swri.edu/index.htm

The purpose of this site is to "honor veterans, living and dead, who served their country on either side of the conflict." It provides an interactive, online forum for Vietnam veterans and their families and friends to exchange information, stories, poems, songs, art, pictures, and experiences in any publishable form.

Radio Talk Show

POW/MIA Freedom Radio American Freedom Network
Station: 1360 AM or 1370 AM
URL: http://pages.prodigy.com/powforum/radio.htm

This talk show hosted by Darwin Rutz is broadcast from Johnstown, Colorado, on Sundays (mountain time) from 3 to 5 P.M. Rutz interviews speakers such as family of POWs and MIAs, activists, and government officials on a variety of related topics. This show can be heard outside the Denver area by satellite rebroadcast or by means of the Internet. To tune in by computer, you need a sound card and speakers.

Videotapes

Americans Abandoned
Type: VHS
Length: 58 min.
Cost: $19.95 plus shipping and handling
Source: American Defense Institute
1055 N. Fairfax St.
Second Floor
Alexandria, VA 22314-9544
(703) 519-7000

This video presents arguments for the existence of American prisoners in Laos after the signing of the 1973 Paris Peace Accords that ended U.S. involvement in the Vietnam conflict. It maintains that the Pathet Laos said they held American prisoners, which they wanted to release through a separate peace agreement with the United States. However, the U.S. government, through negotiator Henry Kissinger, insisted that all prisoners in Southeast Asia be returned through North Vietnam, and North Vietnam refused to accept such a responsibility. The video cites the experiences of American prisoners captured by Communist forces during World War II and the Korean War, and French prisoners during the First Indochina War, maintaining that POWs were sent to the Soviet Union and China. Former National Security Agency analyst Jerry Mooney and others assert that during the Vietnam conflict American POWs were interrogated by Soviet intelligence officials and were sent to the Soviet Union. Then the video explores the possibility of a cover-up by U.S. government officials, a "mindset to debunk" among relevant government employees, and faulty analyses made by the Central Identification Laboratory (CILHI). Among those interviewed are former DIA chief General Eugene Tighe; former government employees Bill Hendon, Bill Bell, and Jerry Mooney; military personnel Millard Peck, Mark Smith, Melvin McIntire, and Bo Gritz; MIA family member Marian Shelton; and authors Monika Jensen-Stevenson and William Stevenson, authors of *Kiss the Boys Goodbye* (Toronto: McCleland and Stewart, 1990).

Korean War Series
Type: VHS
Length: 10 hrs.
Cost: $99 plus shipping and handling

Source: AMR
P.O. Box 190393
Anchorage, AK 99519-0393
(907) 357-7300
Web site: http://adhostnt.com/amr1/

This five-volume set chronicles the Korean War from the dividing of the country into North and South, to the precursors of war, to the height of battle, to the final stalemate and truce.

P.O.W.: Americans in Enemy Hands

Type: VHS
Length: 93 min.
Cost: $24.98 plus shipping and handling
Source: Fusion Industries, Inc.
17311 Fusion Way
Country Club Hills, IL 60478
(800) 959-0061
Cost: $19.95 plus shipping and handling
Source: Public Media/Home Vision
4411 N. Ravenswood Ave.
Chicago, IL 60640
(800) 826-3456

This film is the third in the "Trilogy of Tribute" that includes *Return to Iwo Jima* and *The Unknown Soldier* by award-winning producer Arnold Shapiro. Hosted by Robert Wagner, *P.O.W.* combines historical footage with modern interviews as it explores the captivity of nine prisoners of war from World War II, the Korean War, and Vietnam. Surrender, survival, captivity, and torture are among the topics covered in a candid fashion.

The Story of the Vietnam Veterans Memorial

Type: VHS
Length: 40 min.
Cost: $19.98 plus shipping and handling
Source: Fusion Industries, Inc.
17311 Fusion Way
Country Club Hills, IL 60478
(800) 959-0061

Presented from an insider's point of view, this film tells the story of the struggle to build a national Vietnam veterans memorial to honor the 58,000 men and women who died serving their country during the war in Southeast Asia. Included are interviews with Vietnam veteran Jan Scruggs, who spearheaded the effort to build what became "the Wall" in Washington, D.C.; U.S. Senator John Warner; former Senator Charles Mathias; other veterans; and a mother whose son was killed in action. It also includes footage of the Vietnam Women's Memorial, which honors the thousands of women who served their country in Vietnam. The video is narrated by Rock Bleier, a Vietnam veteran and former National Football League player.

Twentieth Century: Korea: The Forgotten War
Type: VHS
Length: 50 min.
Cost: $19.95 plus shipping and handling
Source: AMR
             P.O. Box 190393
             Anchorage, AK 99519-0393
             (907) 357-7300
             Web site: http://adhostnt.com/amr1/

This documentary highlights the course of the Korean War, which began as a United Nations peace-keeping action when the North Korean troops invaded South Korea in 1950. It deals with the military aspects of the war, particularly the strategies developed by General Douglas MacArthur, the supreme commander of American troops.

Twentieth Century: Vietnam: A Soldier's Diary
Type: VHS
Length: 50 min.
Cost: $19.95 plus shipping and handling
Source: AMR
             P.O. Box 190393
             Anchorage, AK 99519-0393
             (907) 357-7300
             Web site: http://adhostnt.com/amr1/

American veterans of the Vietnam conflict describe their precarious existence in Vietnam and how they managed to survive. The optimism felt early in the war about a quick end to the conflict leads to expressions of despair at the death of friends and low morale. Veterans also discuss their experiences upon returning to the United States as losers in an unpopular war.

The Twentieth Century: Volume 2: Air War in Vietnam
Type: VHS
Length: 2 hrs.
Cost: $19.95 plus shipping and handling
Source: AMR
            P.O. Box 190393
            Anchorage, AK 99519-0393
            (907) 357-7300
            Web site: http://adhostnt.com/amr1/

Narrated by Mike Wallace, this documentary describes the development and course of the air war in Vietnam, from the first use of American aircraft to the signing of the Paris Peace Accords after the Christmas bombing of 1972. It includes original television footage of warfare, speeches made by Lyndon Johnson and Richard Nixon, and Communist propaganda film clips of American prisoners of war. Interviews with repatriated POWs, 90 percent of whom were pilots, provide further insight into their inhumane treatment during captivity.

Vietnam: A Television History
Type: VHS
Length: 13 hrs.
Cost: $99 plus shipping and handling
Source: AMR
            P.O. Box 190393
            Anchorage, AK 99519-0393
            (907) 357-7300
            Web site: http://adhostnt.com/amr1/
Cost: $99.98 plus shipping and handling
Source: Fusion Industries, Inc.
            17311 Fusion Way
            Country Club Hills, IL 60478
            (800) 959-0061

This seven-volume set includes much original television footage of the war and civilian activities, as well as retrospective interviews with participants in the historical events treated. It begins by investigating the roots of the conflict in Vietnam and describes the First Indochina War (1946–1954) between Communist Viet Minh soldiers under Ho Chi Minh and the French colonial government that had regained control after the surrender of the Japanese at the end of World War II. After describing U.S. political interests and policy for Southeast Asia, the film treats the outbreak between the Vietnamese and Communist forces in the North and the unstable Vietnamese political regimes in the South. Volume 3 describes the buildup of American forces in Vietnam, providing the example of Operation Cedar Falls, and gives a first-person account of the war. It also demonstrates the point of view of North Vietnamese Communists, along with the opposing views of American soldiers and prisoners of war. Volume 4 focuses on the invasion during the New Year—Tet—in 1968 and how the defeats suffered then by the U.S. and South Vietnamese troops affected U.S. policy, particularly the Vietnamization of the war. Volume 5 deals with the war as it spilled over into Cambodia and Laos and covers the early peace negotiations. Volume 6 describes the opposition to the war and concludes with the Paris Peace Accords in January 1973. Finally, Volume 7 deals with the legacies of the Vietnam conflict.

We Can Keep You Forever
Type:	VHS
Length:	75 min.
Cost:	$19.95 plus $3.95 shipping and handling
Source:	Kultur Video
	195 Hwy. 36
	West Long Branch, NJ 07764
	(800) 458-5887
Cost:	$19.98 plus shipping and handling
Source:	AMR
	P.O. Box 190393
	Anchorage, AK 99519-0393
	(907) 357-7300
	Web site: http://adhostnt.com/amr1/

Investigating whether American prisoners of war were kept in Southeast Asia after the Paris Peace Accords ended U.S. involvement in the Vietnam war, this mid-1980s production by the British Broadcasting Corporation explains the arguments most commonly made during that period. It begins with interviews of several repatriated prisoners, then describes the positions of General Eugene Tighe, Jerry Mooney, Roger Shields, Henry Kissinger, and several people who made live-sighting reports. It also discusses the arguments particular to prisoners possibly captured in Laos by the Pathet Laos.

Wings over Vietnam
Type:	VHS
Length:	6 hrs.

Cost: $49.98 plus shipping and handling
Source: Fusion Industries, Inc.
             17311 Fusion Way
             Country Club Hill, IL 60478
             (800) 959-0061

This three-volume set focuses on the air war in Vietnam, chronicling the largest air missions of the war. It describes the aircraft employed, from helicopters to bombers and spy planes, and the men who flew them.

Glossary

Agent Orange A herbicide widely used during the Vietnam war to defoliate plants in Vietnam. Its contaminant, dioxin, is suspected of causing cancer.

Annam One of three territories that formed the country of Vietnam in 1945.

Army of the Republic of Vietnam (ARVN) South Vietnam's conventional army of 150,000 men created by the government of President Ngo Dinh Diem in the mid-1950s. It swelled to nearly 1 million men during the Vietnam war.

Bao Dai (1913–) Last emperor of the Nguyen Dynasty in Vietnam and puppet chief of state of the Associated State of Vietnam (1949–1955) under French colonial rule. Ngo Dinh Diem defeated Bao Dai in a referendum in 1955.

Central Identification Laboratory, Hawaii (CILHI) U.S. Army laboratory located in Hawaii since 1973 to deal with identifying remains of U.S. service members killed in Southeast Asia during the Vietnam war.

Chiang Kai-shek (1887–1987) Chinese general and nationalist statesman who unified China by military means in the 1920s and led the country against Japanese aggression during World War II. In 1948, when his government fell to the Communists led by Mao Zedong, Chiang Kai-shek settled in Taiwan, becoming president.

Cochin-China One of three territories that formed the country of Vietnam in 1945.

225

Communist cadre A member of the Communist Party whose job it is to instruct others in the tenets of communism.

defector A service member of one military organization who switches allegiance to another organization (e.g., the enemy).

Demilitarized Zone (DMZ) At the conclusion of the Indochina War in 1954, Vietnam was divided along the seventeenth parallel of latitude into North and South, with the DMZ, or area of no military action, between them to prevent military clashes while a political settlement was reached. During the Vietnam war, U.S. officials believed that Vietcong guerrillas infiltrated the South through this zone.

Democratic Kampuchea (1975–1978) Name given to Cambodia during the rule of the Khmer Rouge Communists.

Democratic Republic of Vietnam (DRV) North Vietnam This government declared itself independent of France in 1945, negotiating an end to the Indochina War in 1945. It was one of the signatories of the Paris Peace Accords in 1973. After conquering South Vietnam in 1975, the unified country was renamed the Socialist Republic of Vietnam.

Gulf of Tonkin Resolution (7 August 1964) Resolution in both houses of the U.S. Congress that gave broad authority to the president of the United States to use any level of force in aiding South Vietnam and other U.S. allies in Southeast Asia. President Lyndon B. Johnson used this power to order troops to Vietnam. President Nixon used it to order troops into Laos and Cambodia. In 1970, Congress repealed this resolution.

Hanoi Capital of North Vietnam and after 1975 capital of the Socialist Republic of Vietnam.

Ho Chi Minh (real name Nguyen Tat Thanh; 1890–1969) Vietnamese Communist leader who founded the Indochinese Communist Party in 1930. He led his people, the Viet Minh, in a struggle for freedom from the Japanese during World War II, from French colonial rule during the Indochina War, and against the South Vietnamese during the Vietnam war.

Ho Chi Minh Trail A network of trails running north-south through the jungles of Laos and Cambodia.

Indochina Eastern part of the large peninsula south of China. It includes Vietnam, Laos, Cambodia, and Thailand.

Joint Casualty Resolution Center (JCRC) Formed in 1973, this U.S. military organization was tasked with determining the status of U.S. military and civilian personnel who were unaccounted for because of the war in Vietnam. It was replaced in 1991 by the Joint Task Force–Full Accounting.

Joint Task Force–Full Accounting (JFT-FA) Replacing the Joint Casualty

Resolution Center in 1991, JFT-FA deals directly with accounting for American MIAs. It has offices in Hanoi; Bangkok, Thailand; Vientiane, Laos; and Phnom Penh, Cambodia.

Khmer Rouge Cambodian Communist group under the leadership of Pol Pot that overthrew the Lon Nol government of Cambodia in 1975. The Khmer Rouge systematically killed some 2 million educated and middle-class Cambodians. When Vietnam invaded Cambodia in 1979, it drove the Khmer Rouge into Thailand. The Khmer Rouge fought the Vietnamese-backed Cambodian government. In 1989, Vietnam withdrew its troops from Cambodia, and United Nations–sponsored elections took place in 1993. Prince Sihanouk returned to lead for a short time a coalition government.

KIA Killed in action.

KIA/BNR Killed in action, body not recovered.

Kissinger, Henry (1923–) President Nixon's advisor on national security who negotiated the Paris Peace Accords that ended the war in Vietnam in 1973.

Le Duc Tho (1911–) Vietnamese Communist statesman who negotiated the Paris Peace Accords that ended the Vietnam war in 1973.

Lon Nol (1913–1985) Right-wing Cambodian lieutenant general who overthrew Prince Sihanouk in 1970. In early 1975 before the Communists led by Prince Sihanouk and Pol Pot took over, Lon Nol escaped to Indonesia.

Mao Zedong (1893–1976) Chinese Communist leader, founder of the Chinese Communist Party (1921) and the People's Republic of China in 1949.

Montagnards Tribal people living in the mountainous region of central Vietnam.

National Liberation Front (NLF) Founded in 1960, this organization coordinated efforts to overthrow the Ngo Dinh Diem regime controlling South Vietnam.

Ngo Dinh Diem (1901–1963) The president of South Vietnam (1955–1963). After serving as prime minister in 1954, Diem overthrew Emperor Bao Dai and named himself president of the republic in 1955. Instead of holding elections to join North and South Vietnam as mandated by the 1954 Geneva Convention, he ruled in a dictatorial fashion. He was assassinated during a military coup in 1963.

Nguyen Van Thieu (1923–) Dictatorial president of South Vietnam (1967–1975), having led a coup that toppled South Vietnamese president Ngo Dinh Diem. A staunch anti-Communist, Thieu was president of South Vietnam during the Vietnam conflict, resigning in 1975 and living thereafter in exile.

Operation Homecoming Name given by the Nixon administration to the negotiated release of 653 American POWs, who were returned to U.S. custody between 27 January 1973 and 14 April 1973.

Paris Peace Accords Familiar short name given to "The Agreement on Ending the War and Restoring Peace in Viet-Nam." It was signed by the Democratic Republic of Vietnam (DRV), the Republic of Vietnam (RVN), the Provisional Revolutionary Government of South Vietnam (PRG), and the United States on 2 January 1973.

Pathet Lao Communist Laotians who aligned themselves with North Vietnam during the Vietnam war. In 1975, this group took control of the country, establishing the Lao People's Democratic Republic.

People's Army of Vietnam (PAVN) Army of the Socialist Republic of Vietnam (SRV), founded in 1945, which fought against the French during the Indochina War (1945–1951) and the South during the Vietnam conflict. In 1978, this army invaded Cambodia.

People's Republic of Kampuchea (PRK) Name given to Cambodia by the Vietnamese-led government in control after the invasion of Cambodia by the Vietnamese (1978–1989). This government was not recognized as legal by most other governments of the world.

Pham Van Dong (1906–) Prime minister of North Vietnam (1955–1976) and of the Socialist Republic of Vietnam (1976–1986), he was the primary leader of North Vietnam during the Vietnam war.

Phnom Penh Capital of Cambodia.

Pol Pot (1926–1998) Communist Cambodian politician, head of the Khmer Rouge and secretary of the Communist Party of Kampuchea (CPK). He led guerrilla attacks against the Cambodian governments headed by Prince Sihanouk and Lt. Gen. Lon Nol. While in power in Cambodia (1976–1978), Pol Pot's extreme regime slaughtered some 2.5 million Cambodians. After the Army of the Socialist Republic of Vietnam invaded Cambodia in 1978, Pol Pot and his forces were exiled to the outreaches of the country, the Thai border. They continued to oppose Vietnamese rule and in 1993 the government was elected in U.N.-sponsored elections. In 1998, Pol Pot was being sought to face charges of genocide when he was reported to have died.

Provisional Revolutionary Government of South Vietnam (PRG) Created in May 1969 by Communist revolutionary forces active in South Vietnam, this entity was a signatory to the Paris Peace Accords. The PRG ceased to exist after the North conquered Saigon in 1975, its members being absorbed into the newly unified Socialist Republic of Vietnam (SRV).

Royal Lao Armed forces of the Royal Lao government (1947–1975) that were clandestinely supported by the U.S. government. These anti-Communist forces battled for control of Laos in the 1970s and lost.

Saigon Capital of South Vietnam, renamed Ho Chi Minh City after its takeover by the Vietcong.

Sihanouk, Prince Norodom (1923–) Cambodian politician who was head of state (1949–1954, 1960–1969). He tried to maintain a neutral stance during the Vietnam war and was overthrown by Lon Nol in 1970. While in exile in Beijing, China, Sihanouk joined with the Communist Pol Pot. After the two overthrew Lon Nol in 1975, Prince Sihanouk again was Cambodia's head of state, until Pol Pot ousted him. Sihanouk lived in exile in North Korea as Cambodia's head of state. After the withdrawal of Vietnamese troops from Cambodia in 1989, Sihanouk returned in 1993 as the symbolic king but left the country when Hun Sen took over.

Socialist Republic of Vietnam (SRV) Communist state that included what were formerly known as North and South Vietnam before the takeover by northern forces in 1975.

Southeast Asia Eastern part of the large peninsula south of China. It includes Vietnam, Laos, Cambodia, and Thailand.

Tet Offensive Major military campaign launched in January 1968 during the lunar new year's holiday by the Vietcong. While the Vietcong's military gains were not long-lasting, the blow against the morale of American and South Vietnamese troops was.

Tonkin One of three territories that formed the country of Vietnam in 1945.

Vientiane Capital of Laos.

Viet Minh Communist-dominated forces who fought under Ho Chi Minh to gain freedom from the Japanese, who controlled Vietnam during World War II. After Japan surrendered in 1945, the Viet Minh seized power and founded the Democratic Republic of Vietnam (DRV), then fighting against the French and their supporters in the Indochina War (1945–1951). After negotiations to end hostilities in 1954, the Viet Minh controlled North Vietnam. Later the Communists became known as Vietcong.

Vietcong Vietnamese Communists. Collectively the native guerrilla groups that made up the military force of the National Liberation Front of Vietnam. Vietcong was a derogatory name used by U.S. and South Vietnamese troops.

Vietnamization President Nixon's plan, begun in 1969, to gradually withdraw American ground troops from Vietnam so that the 1 million-man ARVN could take over most military operations by 1972.

War Powers Act Passed in November of 1973, overriding a veto by President Nixon, this congressional resolution mandates that the president consult with Congress prior to initiating military actions.

Watergate scandal The investigation of a botched break-in at the Democratic National Committee Headquarters in the Watergate Hotel in Washington, D.C. When President Nixon appeared to be implicated in the illegal activity, he resigned the presidency rather than face impeachment proceedings.

Index

Abrams, Creighton, 10
Accountability, 25, 29, 30
  demand for, 20–23, 27
Advocacy and Intelligence Index
  for Prisoners of War–Missing
  in Action (Internet), 161,
  211–212
Agent Orange, 9, 225
Agreement on Ending the War
  and Restoring Peace in Viet-
  Nam, The (1973). *See* Paris
  Peace Accords
Alcatraz, 11, 198–199, 200
Alvarez, Everett, 9, 196
American Coalition Legion,
  205
American Defense Institute,
  161–162, 200
American Ex-POW Association,
  hoaxes and, 23
American Ex-Prisoners of War,
  Inc., 162–163
  regional chapters of, 163–166
American GI Forum of United
  States, 166
American Legion, 166–170
American POWMIA Coalition,
  170–171
Americans Abandoned (video), 218
American Veterans of World War
  II, Korea and Vietnam
  (AMVETS), 171
Andersonville prison camp, 5
Annam, 225
Antiwar movement, 10, 194
Anton, Frank, 196, 204
  on Code of Conduct, 14

Army of the Republic of Vietnam
  (ARVN), 8, 11, 225
  fortress strategy and, 9
Asian Development Bank, 35

Bang, Le, 57
Bao Dai, 45, 46, 47, 225
  abdication of, 6, 7
Barnes, Scott, 200
Bell, Garnet "Bill," 32, 38
*Beyond Survival: Building on the
  Hard TimesA POW's Inspiring
  Story* (Coffee), 209
Bob Dole's Speech against
  Normalization, 10 July 1995,
  text of, 130–133
Brace, Ernest, 15, 197
Briar Patch, 11
Bright Light, 204
Bush, George, 55, 200
  Executive Order 12812 and, 56
  MIA issue and, 29

Cambodia
  elections in, 35
  U.S. invasion of, 10
Camp Faith, 11
Camp Hope. *See* Son Tay
Canadian POW/MIA Information
  Centre (Internet), 171–172,
  212–213
Carter, Jimmy
  criticism of, 202
  MIA issue and, 4, 27, 73, 197,
  199
Case-Church Amendment (1973),
  24

Casualties, 2, 11–15
American, 70 (table)
Cawthorne, Nigel: MIA families
and, 20
CBS-*New York Times*, MIA poll by,
56
Central Documentation Office, 18
Central Identification Laboratory,
Hawaii (CILHI), 18, 25, 36,
72, 74, 201, 225
MIA issue and, 28
Central Intelligence Agency
(CIA), 4, 204
Central Military Museum
(PAVN), 21, 36, 201
Chiang Kaishek, 45, 46, 225
Childress, Richard T., 38, 207
biographical sketch of, 59–60
Chosin Few, 172–174
Christopher, Warren, 35, 57
Clarke, Douglas L.: MIA status
and, 197
Clinton, Bill
criticism of, 206
POW/MIA issue and, 56
remains repatriation and, 72, 73
speech by, 133–135
trade embargo and, 26, 36, 204
Vietnam and, 2, 35, 57
CochinChina, 225
CodeName Bright Light (Veith), 74
Code of Conduct for Members of
the Armed Forces of the
United States, 74, 203
following, 13–15, 199
text of, 75
Coffee, Gerald, 197
Cold War, MIAs/POWs from, 31,
39, 196
Collins, Donnie K.: on hoaxes,
22
Colvin, Rod, 198
MIA families and, 20
Committee for the Defense of
Political Prisoners in
Vietnam, 174
Committee on National Security,
16, 91, 96, 100
Communist cadre, 226
Complete Text of the Letter from
President Richard Nixon to
Prime Minister Pham Van

Dong, 1 February 1973,
152–154
"Convention for Amelioration of
the Wounded in Time of
War" (1864), 6
Convention for the Amelioration
of the Condition of the
Wounded and Sick in Armed
Forces in the Field (1949),
69
"Convention Relating to the
Treatment of Prisoners of
War" (1929), 6
CooperChurch Amendment
(1971), 24
Cressman, Patrick J., 72
testimony of, 100–106
Cressman, Peter, 72

Daly, James A., 198, 204
Dan Hoi, 11
Daschle, Tom, 37
David, Tom, 205
Defense Intelligence Agency
(DIA), 25, 32, 73, 201, 203,
204, 208
MIA issue and, 18
POWs and, 17–20
Demilitarized Zone (DMZ), 10,
226
Democratic Republic of Vietnam
(DRV), 45, 46, 226
Dengler, Dieter, 33, 198
Denton, Jeremiah, 198–199
Department of Defense (DoD),
POWs/MIAs and, 4, 18
Department of Defense Prisoner
of War/Missing in Action
Office (DPMO), 4, 17, 18, 32,
57, 205, 206, 213
hoaxes and, 22–23
POW/MIA issue and, 56
resolution and, 73–74
Deputy Assistant Secretary of
Defense (POW/MIA), 18
Dewey, A. Peter, 46
Diem, Ngo Dinh, 7, 47, 226
assassination of, 8, 48
Dirty Bird, 11
Dole, Bob: speech by, 130–133
Domino theory, 7
Dornan, Robert, 25, 207

Eisenhower, Dwight D., 7, 47, 48
"Escape and My Darkest Hour"
 (McDaniel), text of, 76–91
Excerpts from the 28 June 1995
 Statement of Judy Coady
 Rainey submitted to the
 Military Personnel
 Subcommittee of the
 Committee on National
 Security, 96–100
Excerpts from the 28 June 1995
 Testimony of Carol Hrdlicka
 at the Hearing before the
 Military Personnel
 Subcommittee of the
 Committee on National
 Security, 91–96
Excerpts from the Statement of
 Colonel Theodore Guy,
 106–113
Excerpts from the Testimony of
 Patrick J. Cressman before
 the Military Personnel
 Subcommittee of the
 Committee on National
 Security, 28 June 1995 at the
 Hearing before the Military
 Personnel Subcommittee of
 the Committee on National
 Security, 100–106
Executive Order 12812 (1992), 56

Families. See Missing in action
 families
Fictional works, 193, 194
Films, 194
Final Interagency Report, 29
Florence prison camp, 5
Ford, Gerald, 53
 MIA issue and, 26–27, 197, 199
Foreign Relations Authorization
 Act (1994), trade embargo
 and, 26
Fortress strategy, 9
Four Party Joint Military Team
 (FPJMT), 18
Fourth Allied POW Wing, 202
FrancoVietnamese Accords (1946),
 46
Franklin, H. Bruce, 199
Freedom Flight Tribute Page
 (Internet), 213

Freedom of Information Act, 19,
 20
French Foreign Legion, 30

Gaines, Kimball: on DIA, 19
Garwood, Robert, 3, 32
 conviction of, 28, 33, 196, 198,
 199
Geneva Conference (1954), 6–7
Geneva Convention (1864), 5–6
Geneva Convention for the
 Protection of War Victims:
 Armed Forces in the Field,
 The (1949), 6, 72
 text of, 137–147
Gober, Hershel, 35
Griffiths, Ann Mills, 39, 207
 biographical sketch of, 60
 statement by, 113–117
Gritz, Bo, 21, 200
Grotius, Hugo: just wars and, 5
Gulag Archipelago, POWs in, 34
Guy, Theodore, 12, 13, 33, 34, 74,
 96, 205
 statement of, 106–113

Hall, Roger, 16
Hanoi, 226
Hanoi Hilton, 11, 202
Hanoi parade, 197, 199
Hao Lo. See Hanoi Hilton
Harker, David, 204
Hawk's Page (Internet), 213
Heart of Illinois POW/MIA
 Association, 174
Hegdahl, Douglas, 3, 199, 202
Historical Text Archive: Vietnam
 War (Internet), 213–214
Hoaxes, dealing with, 21–23
Ho Chi Minh (Nguyen Tat
 Thanh), 6, 45, 46, 226
 death of, 50
 Geneva Conference and, 7
Ho Chi Minh Trail, 7, 10, 33, 226
Howes, Craig: Code of Conduct
 and, 199
Hrdlicka, Carol: testimony of,
 91–96

Indiana State Veterans Coalition,
 174–175
Indochina, 227

Indochina War (1946–1954), 6–7, 30, 46, 203
"Instruction for the Government of Armies of the United States in the Field" (General Order 100), 5
International Control Commission, 47
International Monetary Fund (IMF), Vietnam loans and, 28, 35
International Red Cross, Geneva Convention and, 5–6

JensenStevenson, Monika: POWs and, 199–200
Jewish War Veterans of the U.S.A., 175
Johnson's Island prison camp, 5
Johnson, Lyndon B., 10, 48
Tonkin Gulf Resolution and, 9, 24, 49
Johnson, Sam, 200
Joint Casualty Resolution Center (JCRC), 18, 28, 227
Joint Chiefs of Staff (JCS), MIA issue and, 18
Joint Economic Commission, 17
Joint Personnel Recovery Center (JPRC), 204
Joint Task ForceFull Accounting (JTFFA), 29, 38–39, 74, 201, 205, 206, 227
JCRC and, 18

Kampuchea. *See* People's Republic of Kampuchea
Keating, Susan Katz, 3, 200–201
Kennedy, John F., 7, 24
assassination of, 48
Diem and, 8
Kerry, John
accounting and, 25
biographical sketch of, 60–61
Khmer Rouge, 52, 53, 227
Killed in action (KIA), 2, 4, 49, 50, 51, 52, 227
Killed in action/body not recovered (KIA/BNR), 3, 4, 23, 72, 227
MIAs and, 2, 27
Kirkwood, R. Cork, 203

Kissinger, Henry, 51, 52, 227
MIA issue and, 26
peace negotiations by, 11, 50
POWs and, 16, 17, 203
Korean/Cold War Family Association of the Missing, 175
Korean Conflict Casualty File, 196
Korean War, 47, 195, 203
MIAs from, 39, 196
POWs from, 31
Korean War MIA/POW Help Desk (Internet), 214
Korean War Project (Internet), 214
Korean War Series (video), 218–219
Korean War Veterans National Museum and Library, 175–176
Ky, Nguyen Cao: election of, 8

Laird, Melvin: POWs and, 15
Lieber, Francis, 5, 6
Lincoln, Abraham: General Order 100 and, 5
Linh, Nguyen Van, 55
Live POW Lobby, 176
Lon Nol, 52, 53, 227
Lord, Winston, 35, 206
speech by, 117–130

McCain, John, 37
biographical sketch of, 61–62
McCain Amendment (1991), 26, 55
McConnell, Malcolm, 4, 16, 201
McCubbin, Hamilton: MIA families and, 20
McDaniel, Eugene "Red," 74, 200
memoirs of, 76–91
McIntire, Melvin, 200
McMillan, Isaiah, 205
on Code of Conduct, 14–15
Mai Lai, massacre at, 49
Mao Zedong, 46, 227
Media coverage, 10, 194, 195
Medical needs, ignoring, 12–13, 14, 201
MIA. *See* Missing in action
Military Assistance Command, 204
Minh, Duong Van, 8, 54
Missing in action (MIA), 3, 4

accounted for (since 1974), 73
   (table)
American, 70 (table), 71 (table)
KIA/BNR and, 2
list of, 2–4
number of, 71
Missing in action (MIA) families
   hoaxes and, 21–23
   travails of, 20–21
Missing in action (MIA) issue,
   xiii–xiv, 1–2, 193, 195
   debate on, 30–34
   literature on, 69
   postwar, 15–30, 35–39
Montagnards, 14, 227
Montesquieu, Charles de:
   prisoners and, 5
Montgomery, Gillespie V.: MIA
   issue and, 25
Montgomery Committee, 24, 27,
   203
POW/MIA issue and, 53
Mooney, Jerry, 32, 200
Morale, 10, 34, 203
Morris, Stephen J., 206
Murkowski, Frank, 37
Mutual Defense Assistance
   Agreement, 47

National Alliance of Families Bits
   and Pieces Newsletter
   (Internet), 215
National Alliance of Families for
   the Return of America's
   Missing Servicemen, The,
   176, 205
National Defense Authorization
   Act (1991), McCain
   Amendment to, 26, 55
National Defense Authorization
   Act (1996), 57
   Section 569 of, 69, 154–157
National Defense Authorization
   Act (1998), 158
National League of Families of
   American Prisoners and
   Missing in Southeast Asia,
   20, 24, 157, 176–177, 200, 207
   Clinton and, 206
   MIA issue and, 18, 28
National Liberation Front (NLF),
   48, 52, 227

See also Vietcong
National POW/MIA Recognition
   Day, 135–137, 158
National Prisoner of War
   Museum, 57
National Security Council (NSC),
   MIA issue and, 18
National Vietnam POW Strike
   Force, 177
National Vietnam Veterans, 205
Needham, Thomas, 206
Nhu, Ngo Dinh: assassination of,
   8
Nixon, Richard M.
   bombing by, 11, 50, 52
   election of, 10, 49, 51
   letter from, 152–154
   MIA issue and, xiii, 26, 197
   POWs and, 15, 16, 17, 203
   resignation of, 53
   War Powers Resolution and, 24
Norman, Geoffrey, 202
Northeast POW/MIA Network,
   177
North Vietnamese Army (NVA), 53

O'Daniel, Larry J., 202
Office for Prisoners of War and
   Missing in Action (DIA
   POW/MIA Office), 17, 206,
   207
   hoaxes and, 22, 23
   MIA issue and, 18–19, 28
   problems with, 19–20
Operation Big Switch, 47
Operation Homecoming, 3, 15, 26,
   33, 52, 197, 202, 203, 206, 228
Operation Just Cause, 177–178
Operation Little Switch, 47
Operation Rolling Thunder, 9
Operation Skyhook II, 21
Operation Smoking Gun
   (computer program), 209–210
Operation Swamp Ranger, 201

Paris Peace Accords (1973), 4,
   16–17, 51, 52, 53, 69, 72, 203,
   208, 228
   Article 8 of, 18
   breaking of, 27
   MIA issue and, 2, 30, 71
   text of, 147–152

Pathet Lao, 52, 53, 54, 71, 72, 228
PBS/P.O.V. Re: Vietnam Stories since the War (Internet), 215
Peck, Millard A., 19, 206
People's Army of Vietnam (PAVN), 228
    Central Military Museum of, 21, 201
People's Republic of Kampuchea (PRK), 54, 226, 228
Perot, Ross, 200
    biographical sketch of, 62–63
Peterson, Douglas "Pete," 26, 57
    biographical sketch of, 63
Phnom Penh, 228
Plantation, 196, 202
Plantation Gardens, 11, 12
Point Lookout prison camp, 5
Pol Pot, 53, 228
Position on United States Relations with Vietnam in the Context of POW/MIA Progress, 26 January 1994 (Griffiths), 113–117
Potsdam Conference (1945), 45
P.O.W.: Americans in Enemy Hands (video), 219
P.O.W. Biography Database (computer program), 210–211
POW bracelets, hoaxes with, 22
POW/MIA Database (Correlated and Uncorrelated Information Relating to Missing Americans in Southeast Asia) (Internet), 215–216
POW/MIA flag, history of, 157–158
POW/MIA FOIA Litigation Account, 178
POW/MIA Forum (Internet), 216
POW/MIA Freedom Radio American Freedom Network (talk show), 217
POW/MIA Interagency Group (IAG), MIA issue and, 18, 28
POW/MIA Task Force, 24
POW Network (Internet), 178, 216
POWs. *See* Prisoners of war
POW's & MIA's Project Interstate, 178–179

President Clinton's Speech on Normalizing Relations, 11 July 1995, text of, 133–135
Presidential Commission on American Missing and Unaccounted for in Southeast Asia, 27
Presidential Decision Directive NSC-8 (1993), 56
Prisoners of war (POWs), xiv, 1, 195
    American, 70 (table)
    international code for, 5–6
    killing, 4–5
    list of, 2–4, 11
    memoirs of, 193, 194
    number of, 71
    release of, 52
    treatment of, 11–15
Proclamation 6818: National POW/MIA Recognition Day, 1995, text of, 135–137
Provisional Revolutionary Government of South Vietnam (PRG), 228

Quang "1205" document, 206

Rainey, Judy Coady: statement by, 96–100
Reagan, Ronald, 200
    MIA issue and, 28, 73, 199
Red Badge of Courage, Inc., 179
Red Book, 201
"Reeducation" camps, 28, 34
Remains repatriation, 15, 29, 30, 31, 39, 72, 73
Republic of Vietnam, 6
Richardson, Elliot: POWs and, 15
Risner, Robinson, 202
Roadmap policy, 29–30, 35, 37, 38
Rockpile, 11
Rolling Thunder, Inc., 179
Roosevelt, Franklin D., 45
Rousseau, Jean-Jacques: prisoners and, 5
Rowan, Stephen A., 203
Royal Lao, 52, 228

Sabog, Mateo, 57
    biographical sketch of, 63–64

Saigon, 228
Sanders, Jim, 31, 32, 33, 203, 206
Sauter, Mark, 31, 32, 33, 203, 206
Schlatter, Joseph A., Jr.: on DIA, 19
Schlesinger, James R.: POWs and,
    15
Schweitzer, Theodore, 201
    discrepancy cases and, 21
    MIA issue and, 36
S.E.A.R.C.H. Inc. (Internet), 180,
    216
Section 569 (National Defense
    Authorization Act), 69
    text of, 154–157
Senja, Jan, 32
Shelton, Charles, 4
Shelton, Marian: MIA families
    and, 21
Shine, Colleen: MIA families and,
    21
Sihanouk, Norodom, 53, 56, 229
Simpson, Alan, 37
60 Minutes, POW issue and,
    199–200
Skelly, Patricia B.: biographical
    sketch of, 64–65
Skid Row, 11
Smith, Mark "Zippo," 200
Smith, Robert C., 25, 37
    biographical sketch of, 65–66
Socialist Republic of Vietnam
    (SRV), 55, 56, 229
    creation of, 54
    MIAs and, 34, 38–39
Solzhenitsyn, Aleksandr: on
    POWs/Gulag, 33–34
Sommer, John F., Jr., 38
Son Tay, raid at, 11, 50, 74
Souphanouvong, President, 54
Southeast Asia, 229
Southeast Asian Refugees, 24
S.S. *Mayaguez,* rescue of, 54
Stafford, Al, 202
Stern, Lewis M., 204
Stockdale, James B., 199, 204
*Story of the Vietnam Veterans
    Memorial, The* (video),
    219–220
Stratton, Alice, 197
Stratton, Richard, 12, 197, 199, 202
    torture of, 13
Strickland, Jim, 204

Support POW/MIA of Austin,
    180

Tap code, 74, 76
Task Force Omega, state groups
    of, 180–182
Task Force Russia, 18, 31
Team Falcon, 21
Tet Offensive (1968), 10, 11, 49,
    229
Thieu, Nguyen Van, 51, 52, 74,
    227
    election of, 8
    resignation of, 54
Tho, Le Duc, 51, 52, 229
    negotiations with, 11, 16
Tighe, Eugene F., Jr., 28, 32, 55
    on DIA, 19
Tonkin, 229
Tonkin Gulf Resolution (1964), 9,
    24, 49, 226
Toon, Malcolm, 207
Torture, 13, 14, 203
Trade embargo, 204
    exceptions to, 30
    lifting, 26, 35, 36, 117–130
Trading with the Enemy Act, 35
Truman, Harry S., 7, 46–47
*Twentieth Century: Korea: The
    Forgotten War* (video),
    220–221
*Twentieth Century: Vietnam: A
    Soldier's Diary* (video), 221
*Twentieth Century: Volume 2: Air
    War in Vietnam, The* (video),
    220

U.S. House Armed Services
    Committee, 25
U.S. House Military Personnel
    Subcommittee, 25
    Cressman testimony to,
        100–106
    Hrdlicka testimony to, 91–96
    Rainey testimony to, 96–100
U.S. House Select Committee on
    Missing Persons in Southeast
    Asia, 24
    on MIA families/hoaxes, 22
U.S. House Subcommittee on
    Asian and Pacific Affairs, 25
    hearings by, 24

U.S. Military Assistance and
Advisory Group, 47
U.S. Ninth Infantry Division,
withdrawal of, 50
U.S. POW/MIAs in Southeast
Asia (Internet), 216–217
U.S.–Russian Joint Commission
on POW/MIAs, 207
U.S.S. *Maddox*, attack on, 8–9
U.S. Senate Foreign Relations
Committee, 25
U.S. Senate Select Committee on
POW/MIAs, 25, 32, 203, 208
DIA POW/MIA Office and, 19
on hoaxes, 23
POW/MIA issue and, 16, 31,
55–56
U.S. War Department General
Order 100, 5
US Veteran Dispatch, The
(periodical), 208

Van Dong, Pham, 226
letter to, 152–154
POWs and, 17, 26
Vattel, Emmerick de, 5
Veith, George J., 38, 74, 204
Vessey, John, 29
biographical sketch of, 66
Veterans Administration, 195
Veterans of Foreign Wars of the
United States (VFW), 37, 182
hoaxes and, 23
Veterans of the Vietnam War,
182–190
Vientiane, 229
Vietcong, 8, 49, 229
attacks by, 9, 10, 11
POWs and, 14
See also National Liberation
Front
Viet Minh, 6, 46, 229
guerrilla attacks by, 7
Vietnam: A Television History
(video), 221–222
Vietnam Combat Veterans, 190
Vietnamese Independence
League, Geneva Conference
and, 7

Vietnamese National Army, 46
Vietnamization, 10, 50, 229
Vietnam Online (Internet), 217
Vietnam's Studies and
Observation Group
(MACVSOG), 204
Vietnam Veterans Home Page
(Internet), 217
Vietnam Veterans Memorial,
dedication of, 55
Vietnam Veterans of America
Foundation, 190–191
VietNow National, 191
Voice of Vietnam, 13

Wall, The (computer program),
211
Wall Street Journal, POWs and, 55
War Powers Act (1973), 24, 229
Watergate scandal, xiii, 53, 230
Watkins, Willie, 204–205
We Can Keep You Forever (video),
222
Westmoreland, William: bombing
and, 9
Wiedemann, Kent: trade embargo
and, 36
Wings Over Vietnam (video),
222–223
Winston Lord's Speech on Lifting
the Trade Embargo, 9
February 1994, text of,
117–130
Wold, James W., 20, 36, 74, 205
biographical sketch of, 66–67
Woodcock Commission, 27, 202
POW/MIA issue and, 54
World Bank, 35
Vietnam loans and, 28
World War II, POWs from, 31

Yalta Conference (1945), 45
Years of Upheaval (Kissinger), 16
Yeltsin, Boris, 32, 207
Young, John, 205

Zalin, Grant, 204
Zoo, 11, 197
Zoo Annex, 11